Sick and Tired of Being Sick and Tired

AN APPROACH TO SUPERCHARGE YOUR SYSTEM AND CREATE BETTER HEALTH AND WELLNESS

Dr. Alan Trites

ISBN: 978-1-7350685-3-4

Dedication

It is with great pride that I dedicate this manual to my wife Angela, our lovely children, my parents who allowed me to beat my own path, and the giants who put me on their shoulders and made this knowledge available to me; Milt, Gordon and David.

Table of Contents

INTRODUCTION

If there was a simple practice you could do to help yourself when you were hurt or sick that could make you better, wouldn't you want to learn it?

Of course, these are conditions not bad enough to warrant a trip to the emergency room. But I want you to help yourself before you need that trip. It is my vision not to let anyone I know or will know become a statistic by not being responsible and taking action for their health. This is only a gateway into igniting a fire in your desire to learn, take action and spread this knowledge. The body is in fact the great healer and we are only the mediators to aid in its speedy recovery.

It is my wish to provide insight to these possibilities, to provide you with ideas for you and your family's care with easy things to do at home, but not be a replacement for sound medical care. Also, to provide education on laboratory tests that would help you with communicating with your doctor. This book is my way of providing a simple educational tool for each of us to gain a role and be the authority in our health.

Decades ago, I got into alternative medicine more as a curiosity to explain how it could work. I was finishing my clinical laboratory science degree and had access to several medical doctors to ask questions to. I chose to ask these doctors because I had sat in on procedures and had a good rapport. I first asked the orthopedic doctors what they thought of chiropractors. The orthopedic doctor only thought chiropractors stretched a joint but couldn't explain how. Then I asked the pain management clinic about acupuncture. The pain clinic doctors said that some people feel better, but there's no science to prove it. Then I asked the endocrinologist and allergist about homeopathy. I was hoping for a better response because they were using hormones and allergy shots and compounds. These therapies were the closest to homeopathy that I could find. Both the endocrinologist and allergist said their process was different but couldn't tell me how or why.

The doctors I interviewed about alternative medicine were not dismissive, nor did they have a negative attitude toward them. What I learned during my inquiry—and I mean this with respect—was that they had no idea about any of these alternative professions. No offense to them.

My experience with medical professionals is that most stay in their lane and don't seek eclectic knowledge. We have patient after patient come into our office and say, "Why doesn't my M.D. know that?" In reference to nutrition, diet and exercise. The skinny is they have, even at the most prestigious medical institutions, about one hour of education. This isn't their forte nor is mine medicine.

For me, I started some deep research of my own. There are actually thousands of peer-reviewed published research papers on alternative medicine listed on PubMed. This is a medical publication website run by the United States National Library of Medicine. It has strict rules and regulations about how and what gets published. These rules include peer-reviewed papers, settings, sample size, types of studies and outcomes. In other words, you can't submit a paper or literature review without a great deal of effort.

Blood tests and Heart Rate Variability tests are very hard if not impossible to cheat. I have benefitted in my education learning functional medicine, functional nutrition, functional neurology, genetics, methyl genetics, and epigenetic expression to name a few. I discovered there were three variables that came along with pathology. These exist regardless of condition or disease. You have to address inflammation, swelling and blood sugar. It didn't matter the type of doctor or technique. Most doctors were addressing at best one variable regardless of where they attended school or what letters they had academically earned. When I figured out I could address all three variables, I learned something profound. I learned to do a great job at preventing disease and minimizing it in those who already have it.

Medicine—or more specifically emergency care medicine—is wonderful and lifesaving, but where's the prevention? Yes, accidents do happen. But the majority of cardiovascular disease, obesity, diabetes and many chronic illnesses are preventable. Our nation due to our lack of education for health, how to maintain health, and what starves us of our health makes us number 1 in spending and 37th in world health.[1] [2]

In general, Americans go to the doctor after a problem has occurred. Isn't this too late? Wouldn't you want to enhance health outcomes and promote prevention? In a perfect world, the combination of medicine and natural medicine would be an ideal solution. I'm a realist. There are conditions that are so far degenerated, or idiopathic—meaning no one knows why or how it occurs—that medical intervention is the only reasonable course of action.

Hippocrates once said that your food be your medicine and your medicine be your food. We will dive deep into this in future chapters. Your health is interconnected to many other processes in life. To erect a home, one must have good building materials and a solid foundation. If you have a good foundation but lousy materials, then the home won't be that sturdy. Or if you have great materials on quicksand, you're also out of luck. I say you need both.

You are the best person to be your primary care provider. You are the most qualified with life experiences on the beautiful being that is you. You are the one who's responsible to assess your life and to live a life of wellness. When people ignore their health, so many other facets of their life suffer.

You can have physical pain or joint pain that decreases your movement, or your ability to do things with your family or friends. You might have a mental or emotional pain that changes your social output, interactions, energy levels, and drive. And you can have chemical pain that makes it difficult to go out to eat, choose a perfume, or make food for people.

I don't like the cliché that pain is weakness leaving your body when I can appreciate pain as a warning sign.

Once you understand it yourself, then you can make the changes in the energetic systems of your body. You can address the three pillars of health.

Many of the topics in this book are from the most common symptoms and conditions observed. I took my experience through two decades of hospital employment and private practice. In 2007, I spread my practices throughout the country and it became impossible to see patients immediately when they desired. When they had recurring symptoms and I couldn't see them for a month, what could they do?

The true definition of doctor is teacher.

It was then I began telemedicine, in 2005. To help the lives of many patients in search of relief, I needed a go between until they could seek care. I had to teach them what to treat because I may not see them for 6-8 weeks with both of our schedules. It was then my purpose was to create a self-help manual and an explanatory book to simplify treatment so patients could treat themselves until they saw me or their doctor. This became my mantra and this is part two. Enjoy and Be well,

Dr. Alan Trites

"We make a living by what we get, but we make a life by what we give." Winston Churchill

CHAPTER 1

To Function and Thrive you Must Make Energy!

What is fatigue and what is being tired? What does chronic fatigue mean to someone? For some people, being fatigued or tired every minute of every day can ruin their lives. It can certainly lower their quality of life. For some, being tired will cause them to lose their job, or ruin a relationship. And it certainly can limit or impact your ability to finish tasks and maintain your responsibilities. In the United States about 75% of people are tired all the time. In reading this, chances are you are whom I am speaking to right now. Does this mean you can't function? Not necessarily, but I want to help you function better. This is more than taking an energy booster or drinking caffeine throughout the day. Or better yet, taking something that can give you energy that lasts up to 5 hours. But being tired is more complicated than just taking an energy boost drink. If you didn't know, more than 60% of Americans wake up tired and they don't feel rested. Ugh! You were able to get the necessary sleep that your body needed, and you still feel tired. How frustrating! Two-thirds of all Americans are worn out by the end of the day and almost half of Americans are too tired to function properly. To make matters worse, 20% of Americans are sleepy throughout their workday which unfortunately can lead to work accidents or mistakes.

Being tired is a big deal. The vending machine and most coffee brewers know you're tired. They fill their products full of sugar, salt and fat to create a quick, but unsustainable energy. To make matters worse, those foods and ingredients eventually cause your fatigue and to a degree, own you. When you have to reach for those vices to make energy, you have a net loss leading to more fatigue and tiredness. You steal from health, and you become an addict.

In a clinical setting, we can actually decipher, or quantify, how tired you really are. I understand we all have ups and downs, and we all have deadlines or alarm clocks, but maybe you feel you are more tired than the people around you. Here are a few questions you can ask yourself:

What is your motivation when you're tired and are you able to continue doing things?

If you exercise, do you crash?

If you don't do anything, are you still tired?

When you are tired, does it interfere with your physical function and being?

Are you constantly thinking about how tired you are?

Does your fatigue interfere with your work or responsibilities?

Do you feel that you're so tired that it is disabling?

If you're sleepy then how do you stay awake, what do you do?

Are you sleepy all the time?

Are you so tired that it interferes with your work or social life? (Meaning you stop being social.)

Do you fall asleep when you:

Sit to read or watch TV

Are a passenger in a car or worse when you're driving a car

Sit in traffic

Sit after lunch and you're sober

Are in an active place, like a theater, business meeting, or out to dinner

Talking with people in a relaxed phase?

If you answered yes to any of the previous questions, your body is not making energy efficiently - and it's a main reason for your tiredness. If you answered yes to several then your lack of energy is a bit more serious.

Some other symptoms of being tired include:

Being less alert

Difficulty paying attention

You have to pace yourself throughout the day

You become clumsy

When it comes to physical effort - no thank you

You have started to avoid social activities to spare energy

You are more forgetful (your brain requires energy in)

You can't seem to finish tasks

Your thinking and concentration are diminished

Your muscles feel weak

It's hard to make decisions

There's a myriad of reasons that cause you to be tired. It could be diet, lack of exercise or too much exercise, metabolic issues or blood sugar handling. It could be a little bit more complicated such as inflammatory conditions. You could be dealing with blood sugar conditions such as diabetes or hypoglycemia (low blood sugar). There is the possibility of food allergies, autoimmunity, and blood supply issues along with neuroinflammation, anxiety, or depression. There are also permeable membrane issues such as leaky gut, leaky brain, or leaky lung.

You can also have diseases that are taking your energy. If you've ever been sick, that's a pretty good reason why you're tired. It is because you are fighting an infection, or more specific a clinical disease that you will overcome. Right now, you need to sleep to heal.

When you have multiple causes for fatigue, pain and other conditions impeding your health, you need to be tested. Why? Are there any red flags for a serious illness? Missing these can be detrimental and ignoring these could be deadly. Are there any signs of current or past infection? Do you have a thyroid hormone imbalance? Do you have any autoimmunity? Do you have any previous concussions and brain trauma? How is your digestion? Do you have any known or symptoms of allergies? There's a lot to consider. And a good reason to have tests run.

Really these conditions are just symptoms. They are at the fundamental state of how your body makes energy. For instance, if you have one person in your family that is very efficient at making energy and another family member who is not, they will respond to life much differently. The relative that has an efficient energy system will have a healthier brain and they will be more active. They will recover quicker from viruses and other illnesses and their risk for diseases is less. They are more likely to experience more pain, they will more often have mood conditions such as anxiety and depression. Their brain degenerates faster so their memory, cognition, and motivation are all lowered. When you have less energy, you have a higher risk for cancer, cardiovascular disease and a host of other inflammatory conditions, illnesses and diseases. Please tell me you get this by nodding your head.

Here are some big signs you are not making energy. Are you listening? If you've been in an accident where you haven't been able to recover. Or you are depressed or anxious all the time. When you exercise, you hurt for 4 days and can't recover. You have had a virus and you still feel like you haven't recovered. You can't get going in the morning or your muscles are tight all the time. You have constant brain fog or your entire body hurts (like fibromyalgia). All these previous statements mean you have a very inefficient energy system.

I'll take one clinical example and put some of this into context. If you're around a chemical and you get a headache, or fatigue or, or you still feel lousy after exposure, there is a reason for this. You have a very inefficient energy system, and your detoxification can't get enough energy to lessen your burden and now you have symptoms of chemical exposure. This is how one person can react significantly to a perfume or gasoline, and the next person smells it and has zero reaction and looks at you questioning what's going on inside of you. Too many doctors shrug their shoulders to these patients or suggest they see a shrink. While some cases require this, the majority do not.

So, the end game and your ultimate goal is to figure out how to make energy. The definition of making energy in the body is called bioenergetics. This is not energy medicine or philosophical medicine or quantum science medicine. Bioenergetics is how the cells create energy. It's a real term that encompasses what is called cellular respiration. Many of us have heard of something called a Krebs cycle. It can run anaerobic (with oxygen) or aerobic (without oxygen). So back to Biology 101 you must go. When this occurs, it is in the powerhouse in your cell called the mitochondria. The mitochondria is where a molecule called ATP is created. Every system in your body requires ATP. Why aren't people looking at this? I wish I had an answer for everyone, but to tell you a family secret- I havebeen able to look at complex things my entire life and figure out a fundamental solution.

CHAPTER 2

I Was Once Like You

I was once a broke college student with two degrees. One in molecular and cell biology with a minor in chemistry, and another in human anatomy and physiology. My job outlook was $10 an hour in my field. At that time the McDonalds in my town was paying more. I was broke, working two jobs, one for $5 an hour at a pet food store checking pH in water for fish, and the other working for a pet food manufacturer testing food for proper labels such as fat, carbohydrates, and protein. The reason I chose these fields was because I wasn't the typical healthy kid. Sure, I didn't have a childhood disease that was disabling, but I had a lot of struggles. After being adopted, my family couldn't figure out what food I could eat because I was having allergic reactions to everything. I didn't walk until I was 3 and I didn't talk until I was almost 4. When I was 8, I had a traumatic brain injury after being struck by a car while riding my bike. After that day, I didn't move much. I sat around, ate, watched TV, and read books. (It helped that my mother was a librarian, so I had instant and unlimited access to information via the Dewey Decimal System.) In the early 80's sitting around watching TV was not much different from today. At one time in 6th grade, I had 27 chins. In 8th grade after suffering what would have been in today's world a concussion during a football game, I decided to focus on running cross country, soccer, and track and field instead.

There is a saying that you can't turn a pig into a racehorse. Well, my genetics certainly limited my athletic ability. But I do have drive. I knew how to set goals and work hard.

In high school before most people knew how to train, I was waking early and running before school. I was running after school and doing other exercises after practice. I learned how to train my body. I learned what worked for me and what produced results. Mind you this was pre iPods for running. And while I had a Walkman, the batteries were out of my limited budget. Nah. I could do math calculations or prep for the next race during my runs. And many of those runs between ages 15-25 covered a decent distance averaging 16 miles a day. I graduated to longer track races, marathons, triathlons, road races, ultramarathons, and adventure racing. I learned how to make a pig run very fast and sometimes beat a racehorse. I did this seven days a week, week in and week out.

As I began college, I had some life challenges put in front of me and simply applied what I did with running to accomplish them. I was in college and needed to catch up on multiple degrees. I needed to take 20+ credit hours a semester. I need money. I can do school and work part time. My wife (at the time) and I had a baby. I can do school and work nights full time.

I went to chiropractic school and took acupuncture and kinesiology classes in addition to school. I can do that and take 30 credit hours, clinic, full time and work out. Who needs sleep?

When I graduated, I started a practice, and then another and another. People wanted to see a doctor doing different things and getting results. But I didn't have the income to pay for all the help a multi practice needs to thrive. I had to put on a doctor hat, a marketer hat, an admin hat, HR hat, business advisor hat, accountant hat and legal hat. I worked 80 hours a week and donated time on weekends to coaching and more education. Then, to add to things that are awesome but also take time and money, I had more babies.

And then, I crashed. And I crashed hard.

And as a male, I did NOTHING to counteract it. It will pass, right? I didn't change a thing because I could still work. No one in the history of time has ever done this right?

Years passed and I managed. Then it got worse. I have a gift where I can memorize numbers and names very well. And then, I couldn't. I couldn't perform my job. I couldn't support my family. I couldn't be me. Life pushed back. It ended a marriage. It ended all but a percentage of one practice. It ended my desire to run. I was done.

Was it genetics? Was it some bizarre disease? Was it a parasite or yeast? Could it be allergies, food or the environment? Perhaps I had heavy metals, something many alternative practitioners say is a main cause of fatigue and disease. Perhaps I had Lyme's disease. I went to a lot of doctors. I spent a lot of money that I didn't have to spend. Some helped me feel better but generally for only a few days or weeks. (No one has ever experienced this right?) I knew there had to be more.

Then, after asking seemingly everyone to help me achieve a resolution, I had to personally get involved. Fundamentally you have to make energy to function, to have an immune system, to digest, to detox and making energy is paramount! No doctor I visited looked at WHO I am. What I had done (to myself). I had a dozen diagnoses and took anything and everything. I did anything and everything without results.

Then to be told "perhaps it's in your head." Bull butter! You know how you feel, and no one should be allowed to tell you differently how you feel. You know when something isn't right. I can accept I burnt the candles at both ends and candles who were not my own. I can accept the consequences. But to not be able to function at any capacity and being told I'm making this us is complete nonsense.

I'm Dr. Alan Trites. I have fatigue, allergies, autoimmunity, digestive and gut dysfunction and a multitude of concussions and brain trauma. I've been so beat up and tired I couldn't get out of bed for weeks. I wish this on no one. For cases like these there's a lot of "what-if's" and many different disease names. We can test all day long for how you came to be in a doctor's office. The missing factor is who you are. How do you respond to exercise- food- nutrients and what makes you who you are? Without that info how is anyone going to help you make energy? And without energy, you are lost in your quest for health?

My patients have autoimmunity, infections, fatigue, hormone imbalances, migraines, allergies, brain fog and other complex ailments. How am I going to help them across such a wide range of problems? I knew blood work from an additional degree in clinical laboratory science while working in stat labs next to the ER during the night shift.

I asked myself, ``Is there a way to correlate blood work to nutrition, diet, exercise?" Before the turn of the century, I began learning about integrative medicine for doctors who use terms like root cause, epigenetic expression, and holistic health. They were really talking about function. Now, some might call this functional medicine or alternative medicine but that is a name like a shoe company. What does that mean? Do you know something different than other doctors? Are you doing something different than other doctors? There's a big difference between a SAAS and Christian Louboutin, but they're both shoes. There's a big difference and no liability if someone touts functional medicine or alternative medicine. They are indeed keywords, but what are those doctors doing? I'm telling you this as a patient, who like most of my patients have spent $30-40k out of pocket with little to no results. **It's frustrating**. I'd rather have a plan and understand the why of who I am, than a guess of who I am.

Again, I know I'm a race pig; and humor me that I *used* to be a fast race pig. And I prefer to be one again. Here's what I do know. I may not be a pioneer- a visionary- an inventor but I am a good learner. It's why I have spent well over $1 million in education. Why? Because everyone's different. There are an awful lot of different people out there with different how's, whys and whens. I need to do something for someone who wants to get better. But sometimes someone tells me they have allotted $37 to see me with a Starbucks in one hand and their $1000 cell phone in the other. Well, nuts. For them there's a lot of pathways to investigate for someone who is tired all the time. Perhaps you have chronic fatigue syndrome, autoimmune, Epstein Barr, blood sugar dysfunction, have allergies or a disease. Perhaps you have medications or a diet that is affecting you. Or the last time you exercised was last century. I get it, life happens. What can you do to change your course?

My desire was I wanted to be able to function. I needed answers and a solution immediately. I needed to be there for my kids. I needed to be there for my patients. What if someone came into my office like this? How on earth could I help them? I had family members who had cancer, heart disease, diabetes, and other autoimmune diseases. How did these occur? Can we stop them? Is there prevention?

Was it genetic, was it environmental, was it something they did or was it something they ate? More importantly, how were they being treated and what prognosis our outcome was given? It was then I learned a common denominator. Diet and nutrition, according to many of the doctor's opinions in those hospitals I worked in, was not a priority and was clear in the hospital food provided to ill patients. Exercise was recommended but not required. Many doctors saw the same patients over and over again for the same problems. And both patients and doctors were frustrated with the lack of progress or resolution. Was our health care that ineffective? I think you can tell by our country's experience with the latest pandemic it's been bad for a while. I had to do more. Not to mention we are barely in the top 50 countries in terms of "health." As a culture we suck at helping tired people. This includes tired people leading to chronic conditions. Without energy you can't heal. If you have two people one with energy and one without- who heals faster? Can you think of a current pandemic where one person who couldn't detoxify, or had a chronic condition didn't fare well? Or they didn't fare well and found they had chronic conditions. Compared to a healthier person. Let's say they contract a virus to not be named. And while they were sick, it was for about two days and then they are skipping around again. It comes down to energy.

My External Struggle:

The thing is I was fed up at this point with doctors. I was really struggling with treating myself as a patient. But I had to gain energy to work. No energy means no income. I felt terrible because at that point I had 5 degrees and spent $700k in education. Now I was thinking I'm never going to pay this back and I won't be able to help myself or anyone else. I felt even worse because I couldn't support my kids. I was no longer paying my student loan and it was in deferment, credit cards were going up and to try to make things better I took on extra jobs in several different states. (I know you've probably had a bad situation where you did what you could to help but made things worse.) In my office I was struggling to work. I went from knowing names, phone numbers and the names of their children and pets to saying "hey you" when greeting them.

I've been there, I've really felt like crud before

At times there were good days and bad days. Thank goodness for the good days because I had money for food and shelter. I felt terrible because I slowed down my practice to a crawl. I had no stamina physically and I couldn't think for long periods of time. It was a double whammy, physical and mental. Then, as if I was recently adopted, I started reacting to all sorts of foods, chemicals, and my environment. I started to question myself as a doctor. If I can't fix myself, how can I help someone? I can't run with my kids at the playground, what good am I? But now, I can't afford presents or spend time with kids. What kind of dad am I?

When you're tired, mentally and physically, and continually pressing to exhaustion, it puts a strain in one relationship. In my personal story, **one Thanksgiving I came home after a work trip to divorce papers.** It was ugly and drawn out. This went on way too long for two broke people.

I felt like a failure. I felt helpless. It was devastating physically and mentally. All this, while my health continued to deteriorate. As legal fees and actions were adding up, my car was repossessed. I had to move back with my parents and borrow their vehicle. I didn't want to work in healthcare anymore. I looked for jobs in sales, computer work and did odd jobs when I could.

Good news: With all my free time I could sit around in comfy clothes waiting on a dial up modem to work. Meaning, while in my parent's basement I dug deep into my education and tried to figure out the details of why I was so tired. For me this point in my life was rock bottom. No job, no kids, no money, and lots of failures. I was sick and tired of being flipping sick and tired!

I began running tests and what I found was that I was allergic to most foods. The reason, a permeable membrane or leaky gut. Which was caused by either my head traumas, medications, or the active EBV that I found or, possibly all of them. Any could be the cause of my autoimmune diseases SAPHO, DISH and Hashimotos. Which was the cause of my brain fog. Not one of the doctors that I'd spent thousands of dollars with had tested. Or even said those were a possibility. Yet in my parent's basement on a dial up modem, I found the research supporting my conclusion in peer reviewed medical literature.

Then, as if by chance, something amazing happened...

I had an Epiphany!

That's when I found solutions in literature. There were dietary guidelines, exercises guidelines and lifestyle changes to help me feel better. There was genetic testing for function and lifestyle changes to get autoimmunity into remission. Why hadn't any of my doctors, medical or alternative, been talking to me about this?

I brought my updated tests to the medical world. And I was diagnosed with several more autoimmune diseases with many systems in distress. I was prescribed medications and recommended to the Mayo clinic and decided it was in my best interest.

I did as I was told. And being insane (defined as doing the same thing and getting the same results but expecting something different) I was told there was nothing that could be done for me. Try meds and if this one doesn't work, there is another to "try." And good luck. And like that, I was on my own.

Frustrated and defeated, I got a little mad. I said "You know what? Screw it. No one has helped me. No one has a solution or even an offer to help me function better. I'm going to follow what the literature is saying. And I'm going to tailor what that says to my health."

The reason why I was frustrated is that we are so multifactorial. And it didn't matter what doctor you were at, they can't know everything. The only way to have a quality plan of success is to know all the variables. I need to know the lives of my patients. Those doctors didn't know me. I did. They didn't spend more than 15 minutes with me at a time. How could they know me?

I needed to know more and look at my history. Then I could apply it to others. A doctor should ask some or all of the following questions about you. How did you grow up? Were they ever sick or left the country? Any traumatic events? Family history? Genetic history? How do you eat? What do you eat? What type of exercise do you do, if any? What have you gone through to end up in this office?

After tailoring the research for me, my autoimmune cases were in remission. Again, once you have an autoimmune disease you have it for life. Remission means it's not running your life. I'm back to working out (cautiously but with the guidelines). I continue to run tests for improvement. I've had to change my lifestyle, but it's personal to me and it's something I can live with.

I realized that the secret is that the body could heal itself, given the right environment. That environment the way I see it is a balance of stress, exercise, and nutrition. There are *times medications are warranted, there's no disagreement with that, but they don't grow on trees.*

What are we missing? What are you missing?

CHAPTER 3

It's Time to Rethink, Regroup, and Revitalize Myself

I started writing about how we eat, how to shop, and collecting recipes. My medical research gave insight into when you should consider what diet and when. They include anti-inflammatory, paleo, ketogenic, or intermittent fasting dietary plans. Which one is best for longevity, weight loss, energy, hormone, brain function etc.? But I did not stop there.

I learned it myself, lived it, and began teaching it to my patients. We test to make sure they qualify for their health and where we are going. We make a plan for success with nutrition and supplementation. Understanding their genetics, their environments and stress relief. And we test along the way. I had to learn how to apply this to me so I could create it for others. I began practicing again.

The biggest mistake I made in myself and in my practice was not retesting when my patients were doing so well. Even with the pandemic the updated guidelines are to have blood tested twice a year. The tests need to include inflammatory and blood sugar markers. These are hallmarks for disease indications. In the past I may have only tested once a year like most doctors. Now we make sure we are at six months because every cell in your body except your nerves turn over by then. They are no longer the cells we had tested before. How can you prevent it if you don't know and test? I made that mistake on myself and my patients to save money. That model doesn't work. Have you seen the chronic state of America? But that's now how the human body works. It changes every six months. *But there was still a problem...*

The Conflict:

After learning and implementing all those things for recovery my health was still not perfect. I could talk until I'm blue in the face, send emails and still it can be overwhelming. I know. I've lived it. I do my best to articulate what to do because words like gluten sensitivity, oral or immune intolerance will glaze the eyes over. People tend to get stuck on what it is instead of the solution. That is why I created an educational series for my clients about autoimmunity. What specifics you need to change to be gluten free and dairy free, how to exercise for energy and health. And most important, brain exercises and maintaining your brain. I'm getting better about how to make these easier to understand. Podcasts and our on facebook, instagram and twitter. Forgive me, there is a lot of information in my head. And it's still a learning process on perfect communication to suit everyone's different learning style. But I can promise I'll get there. Here is what I found from the 30,000-foot overview. You must plan food that works for you. You need exercise for energy and a quality life. Our environment and food options keep nearly everyone deficient in nutrients and we need to supplement. You need to stimulate your brain. When you shop for food, prepare a list and stick to it. When and if you travel, prepare a week in advance so you don't sabotage your health. Food prep and planning exercise and nutrition became a part time job! This is why I have additional resources to help you make energy, plan, cook and get better.

Our Big AH-HA and happiness in the office:

I ended up getting so frustrated about planning for meals. There are tons of books and apps but none were following the research. For instance, you can buy a keto book full of dairy recipes or a healthy cookbook full of lectins and nightshades. These will not work. As a family and in my office, we were getting frustrated because there wasn't a cookbook that worked for us, or where to shop or the next step. We decided to make it easier and create a step by step set of videos and resources. The videos make being healthy exponentially easier, and there is a learning curve, but for you it won't be a full-time job like it was for me. When should you eat? Should you do programs like AIP, Keto, intermittent fasting? What does being gluten and dairy free and should you need it in your life? In my experience I have not come across someone who couldn't benefit from gluten and dairy free. Why? Because they are inflammatory. They change your sugar levels. They affect your ability to make energy. If you haven't figured it out yet is the purpose of this book. I talk about gluten and dairy A LOT. There are hundreds of researches showing how it drains you of nutrients specific to you making energy.

If you've ever tried to smoke, hopefully it burned your lungs and throat. You coughed. Your eyes watered. You felt like crud, and you said this is not happening again. You had exposure. You didn't end up with cancer because after a try or two you had already quit.

When you go gluten and dairy free you may not feel anything at all, or you may feel everything is better. You won't know for a month. And I want you to know you can do anything for a month. You are strong enough for that. If you feel nothing, you wouldn't know if your health had improved without some testing or a symptoms survey. How about a brain or reading test and gluten and dairy free can improve brain function? They are inflammatory and inflammation is a root cause of all disease. If you quit this, at least you decreased inflammation for a month. Most importantly, did going gluten and dairy free improve your energy? Did you last longer in a day, need less caffeine and other energy vices?

My Transformation:

I now make energy. I can function. Do I still have autoimmunity? Yup. Once you have it you will have it for life. But they are in remission. Do I still have bad days? Absolutely. The ultimate goal is to minimize those. I learned a valuable lesson. I have to put Myself first with energy. I've discovered what brings me down. Energy suckers come in many forms. It could be food, relationships, sleep, diet changes, medications and other factors. Now, I do whatever I can to not ever be back in bed, living in my parents basement, alone and afraid of the future.

Please enjoy this Sick and Tired of Being Sick and Tired book. Start a goal of improving at least one thing at a time. Change will happen. If you need more guidance watch our videos and get our keto book. And if you need personal guidance, contact our office. You're allowed to be you. My job is to find another way. And now it's fun to be in practice again. I love getting up for work and seeing who we are going to help. I love staying up and continuing to learn and expand my knowledge. I love helping people. I love helping my family.

I've had multi-millionaires in my office tell me they'd give it all to feel good and have energy. They waited and I beg you not to wait. Get educated on what it takes for you to make energy. Our mantra is to help you achieve health with energy, so that you can create wealth and help others.

CHAPTER 4

How You Make Energy- and why it's important to know what pattern you fall into and what may be stealing yours!

You don't breathe or have a heartbeat or make hormones, digest, defecate or do anything without Adenosine TriPhosphate (ATP.) In the presence of oxygen, you make in the mitochondria 36 to 38 units of ATP. When you're sprinting down the road or what's known as an aerobic, you make about 2 units of ATP. All people who are in an inflamed state whether it is in their environment, or diet, or lack of exercise, or a combination of all these plus genetics. All of these make less ATP. In fact, all inflammatory conditions reduce the amount of ATP you can create. So, this means it's between 2 and 36 units of ATP. Either way it's not efficient. The more you fall apart the more you don't recover. And the more you can't sleep because inflammation has an effect on your sleep pathways. Also, let me be clear if you have no movement then your mitochondria can't make energy. So, if you were hoping to find in this book what's the best supplement to take or the best medication for energy production. Or the best food that you should eat to make me feel better and have energy, I don't have it. And the reason I don't have it it's because no such thing exists. Your body is designed to move, that is how it makes energy. We are kinetic and movement creates energy.

Understand that this doesn't mean I need you to power lift. And I'm not asking you to run a marathon. In fact, all I may need you to do is move for 30 seconds at a time and that's it. My job is to bring the research on how to help your body make ATP. Of course, in general terms, until I see you in my office, I cannot create a custom plan for you. If you are not making ATP with efficiency, then you can't recover which means that you can't heal.

This is not a form of aging. This is the mechanism that keeps your body to stay in homeostasis which is normal and heals. If you have chronic conditions such as inflammation, or worse, cardiovascular disease, Alzheimer's, or an illness or autoimmune condition. Then, your lack of making ATP will prolong these conditions and worsen them, or not allow you to recover.

Why aren't doctors focusing on this? I don't know. I am not them and they have their own journey to discover for themselves. You are reading this because those previous avenues have failed. Or you don't want to end up like a loved one or a friend who couldn't make enough energy to be well.

There is no magic formula, or energy drink, or probiotic or miracle diet program. This is much more than a supplement or natural health care. And no, most of the time those once useful B vitamins don't do what they once did. To be frank, you have some issues with energy production that will at some point, if not now, have a negative impact on your health.

They can also be chemical burdens or exposures to an environment whether it's mold or dust or pesticides and other toxins. Chemicals affect you by creating free radicals. Free radicals go to the mitochondria, and they decrease your production of ATP. So sometimes we must look at where you are in life? Where do you live and where are you coming from? Sometimes there are genetics in play which makes you less efficient and that may need to be discovered too. But how do you get out of this? Chemical burdens and free radicals have to be offset with something called antioxidants.

There are few nutrients that do help mitochondrial function and the production of ATP, and they're called cofactors. You could be also missing the proper amount of blood sugar and I can be too high or too low known as hyper or hypo glycemia. Or you may not have enough oxygen to the mitochondria. This is the equivalent of trying to start a fire in a covered area. Without oxygen it doesn't light. In your body that's called anemia. Or there might not be enough ketones to break down fats to get you the energy that the mitochondria needs. Or you might have enough antioxidants as explained before. Sometimes your body doesn't digest well and then you can't digest cofactors. Examples of these are B vitamins or Co Q 10, alpha lipoic acid, carnitine, vitamins adek, vitamin C, magnesium, and calcium. These are the cofactors that help your body make ATP.

This is where a poor diet can wreak havoc on your ability to make energy. When you eat junk food or high-fructose corn syrup or gluten in the form of any grain you get an insulin surge. The insulin surge will damage the mitochondria by way of free radicals and make them less efficient. So, if you're under the assumption that 200 calories are 200 calories let me explain how that is asinine. If you have 200 calories of broccoli which have these cofactors, they allow your body to make energy as efficiently as it can. Hopefully, it will make 36 units of ATP with oxygen. Now, assume you recently ate a 200-calorie fast food or pastry, candy or cola that creates free radicals. Keep in mind that rarely is there 200 calorie junk food. Now your body has to create energy to detoxify. That new rush of sugar from high fructose corn syrup requires energy in the form of insulin. This is to keep you from having too high of an energy or blood sugar spike. Your body will then steal the cofactors that were available for you from other food or your storage. Instead of the food to allow the insulin to do its job. Over time this is how we get to a condition called insulin resistance. You keep challenging your pancreas to make more insulin because of the food choices. And soon, if not already, you will push yourself to the point that the pancreas runs out of insulin, and you are now pre-diabetic. This is how the standard American diet (SAD) pushes us to be sick. If the food we eat makes us sick and the pharmaceutical industry works to fix those solutions, is there ever going to be collaboration to make us healthy as a country?

Or if you recall, there are other inflammatory conditions such as food sensitivities, blood sugar handling issues like diabetes, autoimmune diseases or undiagnosed autoimmunity. You could have irritable bowels, inflammatory bowels, bad digestion, or allergies. Inflammation can come from concussions and other neuroinflammation conditions, anxiety and depression and permeable membrane otherwise known as leaky gut. Again, all information can slow down your ability to make energy.

Unfortunately, the concept of calories being equal is still taught in many schools including those of dietitian and nutritionist. I was taught this in all my undergraduate schools and in clinical laboratory science taught at a medical school.

When you have insulin resistance you have increased blood sugar in the cell, and it takes more acetyl-CoA for the mitochondria to work. You increase what's called reactive oxygen species, which are free radicals. These damage the mitochondria. And then you have decreased mitochondrial biogenesis which is how the whole thing works. Over time, you get systemic mitochondrial dysfunction. When this occurs, this is how your muscles start to ache. This is how you get reduced endurance. I hear phrases such as "I used to be able to and now I can't." Or when I work out even for 10 minutes, I hurt for three days. If it goes to your organs you start to have decreased metabolism. So, this is not somebody who doesn't drink and can have a fatty liver. Or somebody who doesn't eat that bad can have gastritis or inflammatory bowel, or irritable bowel. Or even cardiovascular disease because there's not enough energy to keep the systems efficient. If it goes to your brain you get increased brain fatigue or brain fog. It can also lead to mood disorders whether that's anxiety, depression, ADD, ADHD or other conditions.

To drive a point home, when you have insulin resistance and mitochondrial dysfunction you can get neurodegeneration. This is a word you've heard for the first time that means that your brain and the nervous system are degenerating. Hold on a second oh, we are all degenerating after about two years of age. But how fast you generate depends on mitochondria. And what you do in life to damage it can slow down or speed up the process. For instance, if you don't have insulin resistance and you work out within your limits. This means not too much because over exercise can screw this up too. You will create new connections by something called neuroplasticity. So, as we age we will continue to degenerate. But if we keep our neuroplasticity, we will make new connections and it may not even appear too much degeneration is happening. I want to emphasize it comes down to making energy and how efficient you are in any system.

That you have mitochondria throughout the entire body. It can happen or be reduced anywhere. For many people they have reduced intestinal mitochondria. There are several ways this can occur. You can have a traumatic brain injury, post-traumatic stress disorder, you can deteriorate into a condition called dysautonomia. This means you can't regulate your blood pressure or you're very excited in the near very slow in this leads to almost a fight-or-flight condition. You can have anemia or a pulmonary disease such as you smoked or COPD or reduced cardiac output because you never move leading to systemic hypoxia. This term means there's not enough oxygen getting to places. You can have oxidative stress from exposure to chemicals, toxins, or radiation. Meaning lots of free radical exposure.

You can have intestinal inflammation from an infection such as a virus, bacterial, fungus, mold, yeast, or parasite. You can have dysbiosis from eating too much sugar or antibiotics or a result of an infection. You can have an autoimmune condition that is causing the inflammation. When you have any of these you can have reduced intestinal mitochondria. But what does this mean? Well, when you have reduced mitochondria, you have a loss of ATP or energy. Now, the smooth muscles that help you digest stop being efficient. These smooth muscles create motility, which is how food moves through your digestive system. Without motility this can lead to gas and bloating. It can also lead to lost motility, or impaired motility which is diarrhea and constipation. You could have a decrease in your gallbladder function. You could have the nervous system of the intestines start to lose its function. You could lose the immune system of your intestine. You could reduce or lose your ability to digest by way of digestive enzymes. So this is a Big Deal.

The symptoms lead to hypochloridria, the ability to produce enzymes. And you lose your ability to handle large meals or many meals. Or you are inflamed all the time. Or either way. You know when you eat you don't feel good, and it doesn't matter exactly what you eat it. It could be a bunch of problems or where the inflammation is located based on those symptoms. I'm not here to get into gut physiology. What I'm trying to explain to you is that gut physiology is 100% related to how your body makes energy.

So when it comes to making energy you could have toxins or infections or sometimes physiology that doesn't allow it to happen. This here's where our society suggests as a primary form of treatment is prescribing a pill first. Below you can see how this makes things worse. Below is a brief, and I mean brief list, of medications that actually hinder mitochondrial function and decrease your energy.

Analgesics

Antibiotics

Antipsychotics

Antidepressants

Anti-anxiety

Anti-epileptic

Antidiabetic

Antiviral

General anesthetics

Statin drugs and other cholesterol lowering medications

Sedatives

Pain medications including NSAIDs

You can see by this list how a lot of people are tired in our society because they, or your, have been prescribed these that steal your energy. When we have 60 to 70% of people in this country who are tired. This isn't necessarily induced not from genetics, but from medications prescribed!

CHAPTER 5

Your Big Five: Where Are You Not Making Energy?

Sleep please!

Let's talk about sleep for a minute when it comes to energy. You can have the inability to fall asleep. That could be from psychological stress, or metabolic syndrome such as diabetes too high blood sugar too low blood sugar. It could be a brain injury because you no longer have the ability to slow down the activation system. This comes from a part of your brain called the vestibular or cerebellum, that keeps you lit up and you can't calm down. You can also have hormones such as cortisol, epinephrine or norepinephrine which are excitatory and don't let you sleep.

There are different reasons for how you sleep, such as the inability to fall asleep. Another is staying asleep. Most of the time that is a blood sugar or low blood sugar that causes hypoglycemia at night. You can have another metabolic issue or an increased activation of a particular part of your brain called the limbic system. Or you might wake up with morning fatigue. That could be due to a prostate or a bladder issue. Or you're so pooped out that you can't make a cortisol cell response and it doesn't wake you up. The cause of this could be you aren't getting enough oxygen at night and so you have a condition known as sleep apnea and you need oxygen to help you out in the morning. If it's blood sugar, oxygen or stress, your body figures out by chemical messengers how to live in a fight or flight mode. How much energy reserve do you have? This is how some people can deal with stress and others crash.

These might be the big hitters. There are other conditions that allow you to not sleep at night. If you have fatigue after meals such as insulin resistance, food allergies or histamine intolerance or you have an allergy to MSG. You could have a history of brain injuries or concussions. You might be in neurodegeneration, or there were developmental conditions. Such as you never liked math and you never will do math and so that part of your brain develops slower and now it's affecting you. Or you could not sleep because you're being exposed to chemicals or there's a candle in your room. Chemicals can affect your health regardless if you can smell them or not. You could have new carpet or new paint that's affecting you. When you touch it you don't break out or welt. When you smell carpet, or other chemicals, and you have any symptoms, such as a rash, dizziness or brain fog, this is a loss of chemical intolerance. This means you have a loss of antioxidants which means this creates an inability to make energy to stay asleep.

It could even be antioxidant reserves due to over exercising or you have altered bioenergetics. If you recall, this is the ability to make energy. If you over exercise or don't have antioxidant reserver, this is flat-out brain fatigue because the brain itself is not getting enough fuel. For some, fatigue can be hormonal. It could be too much estrogen or progesterone or testosterone or they are taking these as medications or supplements. There are conditions where one has fibroids, polycystic ovary syndrome, or low testosterone. There are conditions that are pro-inflammatory, and the inflammatory hormones are elevated. Some people can't sleep because they have exposure to pollen or pollution, and they have a loss of what's called their pulmonary barrier. And then they start to make antibodies to it which is an allergy which is a loss of immune tolerance. or when you get exposed to damp buildings you might have a mold sensitivity. Again, all these conditions above reduce your ATP.

Later In this book I'm going to give you some strategies to help with your fatigue and to help your body make energy. But if those simple things don't work, or don't work to your liking. I will always recommend you need professional help. If you don't make energy, you can have an infection, an autoimmune condition, cardiovascular or pulmonary disease, sleep apnea, or diabetes. You may have a metabolic issue such as your thyroid or insulin resistance, or anemia. Any of these conditions deter your ability to make energy. And don't forget there could be drug-induced mitochondrial toxicity from those drugs that are so common. Such as pain medications known as NSAIDs, or statins, or antidepressants, or anti-diabetic. There's a lot. Like I said before this isn't a one-stop-shop for fatigue. But what I have at the end of this book is a survey that you can take. And if it falls into a condition that you have what to do. Do you know how you could make energy better and you're not doing them? Then some simple strategies will help. But I want to be perfectly clear. If you don't feel anything or an improvement, you need to seek professional help somewhere. You cannot ignore this when your body doesn't make energy. And when you are deficient in energy, you don't fight off disease and your expectations of a positive outcome will be limited.

2) Oxygen

One of the biggest things that can slow down your ability to make energy is not having enough oxygen. You might have airborne allergens that can affect you if there is IgE allergy testing to see if you are allergic to it. You can also improve your air quality at work or at home and there are air purification systems or a HEPA air filter. You want to improve your Berry or systems and keep in mind that is your long sinuses brain and gut. Here are some products that help improve these barriers: Vitamin A, vitamin D, vitamin K, and glutathione. If it is a gut issue that is causing your breakdown because they use the same co-factors you might consider short chain fatty acids, fiber, and an anti-inflammatory diet meaning you're not eating foods that inflame your system and use your co-factors. Sometimes it's necessary to consider meeting with an allergist and having low-dose antigens or immunotherapy because they help improve your immune tolerance.

To explain how this works, let's say you have an airborne allergy and it's from pollen or grass or pet dander. These get into the lining within your sinuses or lungs called the mucosal areas. They create inflammation and this leads to narrowing the bronchial tubes. From this you start to get symptoms such as coughing, mostly at night or with exercise or laughter. Then you have difficulty breathing or tightness in the chest or shortness of breath and worse wheezing. And then to make matters worse depending on the weather or food additives, or dietary proteins or even air pollution can stack on top of this airborne condition. As you get an inflammation and airing of your bronchial tubes you get a decrease in oxygen. Now you have a decrease in options to the mitochondria for fuel, and now the mitochondria cannot efficiently make ATP. The result is that you know you are tired. In our office when somebody comes in and they have these conditions, I really need to sit down and listen. When someone says they are tired they could have a multitude of conditions where they're not making enough energy. Over time this means there's probably going to be another issue if we can get ahead of it. But even myself, who has been to plenty of doctors' offices. Recommendations and prescriptions did not work. In any office, whether or not the doctor is listening to your story or not. Not understanding how you make energy or are not, makes it very difficult for whatever treatment or product they use to help you feel better. When your inflammation is stealing the cofactors, you don't make energy. And what a doctor to ask you to take or eat doesn't work very well without cofactors.

What if it isn't an allergy? It could be a deficiency in B vitamins and that can help support adrenal function and stabilize blood sugar. There could be a lack of B vitamins in your diet which means that your adrenals can support your blood sugar up and down. You get too much stress in your life, and you need to reduce your mandates of deadlines and projects or worse you need to reevaluate your relationships. You may have a problem with sleep, and we need to deal with your quality sleep and regular sleep-wake cycle. You can have blood sugar that is too high or too low or you could have some resistance because of your dietary choices. You do not exercise at all which is bad, or you exercise way too much. Or you exercise moderately but you just can't recover.

3) Blood sugar

Let's take a minute to talk about symptoms of blood sugar changes known as dysglycemia. Elevated blood sugar which is known as diabetes if it's elevated too much. Low blood sugar is known as hypoglycemia. Ideally you want normal blood sugar. In this next section, I'll give you some ways to help you self-diagnose your blood sugar and whether it's low, normal or high.

If you have normal blood sugar responses, you wake up with energy in the morning. You wake up with an appetite. You're hungry between meals but you're not craving sugar. Your life doesn't change between meals. And you do not have fatigue or a change in energy after meals. In other words, eating food does nothing to your energy.

If you have low blood sugar, it's very hard to wake up. You typically don't have an appetite or you are hungry in the morning. Hypoglycemics generally start their day with caffeine or sugar. You don't function very well between meals and when you do eat you actually get energy after a meal. Most people crash in the afternoon between 3 and 5 and it could be earlier than that. And when you crash you reach for sugar, salt, or caffeine. Often this leads to difficulty staying asleep throughout the night. Of course, all of those keep you on a blood sugar roller coaster. It's high quickly, and you have energy. And then you crash. Most come into my office and have candy in their bags or purse just in case they get too low.

When you have insulin resistance you can still sleep but wake up not feeling rested or recovered. You will wake up wanting sugar and you typically eat a high sugar or high starch breakfast. There are sugar cravings consistent throughout the day. When you eat you get tired and most of the time it's right after lunch. This is why Starbucks between one and three in the afternoon still has a line like it does in the morning because people need stimulants after their meals. What this also means is you still have difficulty falling asleep.

How do you unwind either one of these? But the idea is that you don't have any simple sugars or sweets such as candy or honey or soda. You don't drink fruit juice. You don't have fruits as a snack or after a meal. And you limit your high-carbohydrate foods such as bread, pasta and rice. Ideally the foods you eat are vegetables that are grown above the ground because they have a higher fiber content. You eat healthy fats such as avocado, olive oil, nuts and other seeds and fatty fish. You eat high protein meals. When you're hypoglycemic you really do not want to miss meals because it makes you more hypoglycemic.

If you tried all of these strategies before and you still don't feel you have any changes in symptoms. If you wake up in the morning sluggish, or you still can't sleep, there is a condition called the Warburg effect. This is when you need to consider getting your body into ketosis. What this means is your body is working more off of fat than carbohydrates. Ketosis can affect an abundance of conditions. It can reduce inflammation, protect the brain, reduce autoimmunity, increase mitochondrial function, reduce free radical and oxidative stress, reduce insulin surges, reduce blood sugar spikes, reduce obesity, seizures and swelling.

Well, we have a ketogenic program and a cookbook in her office. I'll give you some highlights of what ketogenic diet is supposed to be made of. If you need help, call our office. We are well educated in this field and have the ability to cater it to you. It should be about 80% fat, 15% protein and 5% carbohydrate. That is what the research over the past 40 years has shown to be the most effective. Understand that there are some people who don't have a gallbladder that need supplemental aid so that I can get someone into ketosis without falling apart. The permitted foods on a ketogenic diet are meats and seafood and eggs if they are tolerating, you're not allergic to them. All low-glycemic vegetables that are above ground. Oils such as Olive, avocado, coconut and Ghee are better if tolerated.

Foods that need to be in moderation are low glycemic berries are fruits, bacon and salami and not such as pecan, Brazil and Macadamia because they have lower carbohydrates. Bell peppers, brussel sprouts, green beans, tomatoes are a little higher in carbs and you can have small amounts of onions.

Foods that you need to avoid include dairy, except for the butter for those who can tolerate it, processed meats, high-sugar foods, and alcohol.

When you get through ketosis you will start making more efficient energy. And then we go to something called intermittent fasting. Before I get into fasting there are various types of fasting to be aware of. There is a 5/2 fasting which means you eat normal for five days and then for two days you have a 500-calorie day. You could do an eat-stop-eat meaning that you have a 24-hour fast once or twice per week. You could fast every other day. You could consider the Warrior Diet which is a small amount of fruit during the day and just one large meal at night. You could spontaneously just get meals from time to time. You could do a multiple day fasting such as a weekend fast. You can do a fast-mimicking diet which is five days and one of the best out there is prolon. After keto we use time restricted fasts. Options could be a 16 hour fast with an 8-hour window of eating or 12 + 12 or 14 + 10.

The outcome of intermittent fasting is that you have decreased body fat, blood pressure, heart rate, blood sugar and insulin. Metabolically it increases your mitochondrial function and energy production. And increases your recovery and brain function. Fasting increases cancer prevention. It increases your antioxidant production. It reduces neurological degeneration and inflammation.

To be fair if you've gone through ketosis and you stayed with those foods you can always come back to but consider the Autoimmune Paleo Diet. And the foods that are typically permitted on this are vegetables and fruits, fermented foods, healthy oils that you already use a new ketosis and all meats. This diet is to help reduce inflammation and help you feel better which hopefully decreases disease symptoms.

4) Hydration

Of course, if you don't drink any water, you don't have any cellular oxygen for your mitochondria and they will make less energy. The majority of Americans are significantly dehydrated. A symptom can be that you have fatigue which is the number one symptom of dehydration. You can be thirsty or have dry mouth or lips or wake up that way. Or just have frequent urination. When the electrolytes become imbalanced you get muscle cramping, diarrhea or constipation, and increased heart rate and salt cravings. Sometimes you just drink too much coffee, caffeinated drinks, energy drinks or alcohol. All of these are considered diuretics and add to dehydration. Sometimes it's best to set an alarm or have a bottle that shows you how much water you're actually drinking. Most of my patients are surprised how little water they're actually drinking. And sometimes we have to add an electrolyte supplement that has sodium bicarbonate chloride, potassium, calcium and or magnesium.

Sleep is a major factor in how your body makes energy. A few stages in how sleep is inefficient for you or does not. Do you have the ability to fall asleep, to stay asleep, and the ability to get recovered when you wake up? Some simple steps to begin if you're not sleeping well is to limit your caffeine intake later in the day.

Avoid electronic devices before bedtime. Use blue light filters on these devices throughout the day. Avoid exercising too late in the evening. You can always exercise earlier in the day. And then keep your bedroom temperature comfortable, not too hot and not too cold.

If you have difficulty falling asleep there are multiple factors that come into play. It could be psychological stress like a relationship or job. You have a metabolic syndrome such as diabetes. When people have brain injuries and concussions, they can't calm themselves down or when they have conditions that affect their cerebellum, balance, or digestion it also affects our ability to fall asleep. Suggestions to help combat this is to do physical activity throughout the day to help with stress response. To reduce your stress at night the best you can with it is meditation or reading. Avoid overstimulating your brain activity throughout the day and this can be devices or screen time. You can take supplements to help support the Gabba Pathways which slows things down. You can listen to music therapy with Beats less than 60 beats per minute. And if you're on a ketogenic or intermittent fasting diet you may eat healthy fats or take MCT oil or exogenous ketones before you go to sleep.

If you have difficulty staying asleep you typically wake up two to four hours after going to sleep. This can lead to waking up with anxiety or racing heartbeat trembling or shaking. Some of the causes of this or what's called nocturnal hypoglycemia means that your blood sugar drops throughout the night. you can have brain impairment and I won't go into the details of this because you do need a medical history and exam to look into this. Or you might have a metabolic disease such as liver, kidney, pancreas, diseases, or infections. Medications can even create this. How this happens is you can miss meals throughout the day which doesn't allow enough energy to come in or when you have a lack of sleep over time it leads to decreased energy to the brain. The brain requires energy so it's going to create a stress response so that you can get food for the brain, which unfortunately is in the form of adrenaline which creates a fight or flight sympathetic response, and it wakes you up. So, if you're waking up, we want to make sure your dietary concerns are addressed. The other options are weighted blankets because they help reduce your sympathetic response. A white noise machine. You could do mouth taping to help. There are many online outlets where you can buy items to help with mouth taping. Or you can take supplements that help increase your Gaba response. This supplement also helps your brain so down. But you can talk too much or get the wrong type so be careful. Notice I did not say melatonin. It can cause an abundance of issues. If you must, do your best to stay below 3mg that has been studied.

Now, what if you wake up in the morning just pooped out? If you have to use an alarm to get out of bed or the second you get to bed you depend on stimulants to function you really need to consider exercising for function. Sometimes you have neurological movement disorders such as restless leg syndrome. This can lead to poor sleep because you're constantly moving and using energy. Some of the causes of your morning fatigue could be obstructive sleep apnea. This is where there is loud snoring for episodes where they don't breathe for a short time. They wake up with a dry mouth or morning headaches. If this is you then you should do a home sleep test.

If you wake up and have no appetite in the morning. Or awake with anxiety or they have to start moving so that they feel well, this can be an impaired cortisol awakening response. And you need to see a doctor for that. You also have a brain injury or concussion which impairs the brain function and again you see a doctor for that. Now, you might get up at night and go to the bathroom all the time. For men that might be a prostate issue and for both men and women it might be a pelvic floor muscle weakness. You could have a high protein or sodium diet which makes you urinate at night. Or for both you just might have an overactive bladder. For most of these conditions, you cannot supplement your way better. There are strategies in a doctor's office and treatments that are going to be needed outside of the context of this book.

5) Exercise

We need to talk about exercise when you are tired. and before you say I don't know if i can do it, please listen to me for a minute. You can, I believe in you. And your body needs it to get out of fatigue. I'm not asking you to do light exercise which is what most people who have fatigue do. And the definition of light exercise is that you add about half of your max heart rate, and you don't have any noticeable change in breathing or you don't sweat unless it's hot outside. They should be something like going for a walk or walking around the neighborhood. And then there's moderate exercise which you start to sweat after 10 minutes. You notice your breathing is deeper and more frequent and while you can carry on a conversation you can't sing. And there's high intensity. You break a sweat between 3 and 5 minutes and your breathing is deep and rapid. You can only talk in short phases. You can't sing. The first two types of exercise, light and moderate, really don't have any outcome on pain response or how your body makes muscle. It doesn't create nitric oxide meaning oxygen to the rest of your body.

It does increase the ability to have brain function and it has a little effect on your immune system insulin and how your body creates free radicals.

Whether you know it or not, every exercise crates for radicals and you have to have antioxidants in either your body crates and where you supplement them to offset the exercise. We want to avoid those pesky radicals. That's if you're at a conversational pace and you walk for 10 hours a day or you jog for 30 miles you will create oxidative stress. There's a thing called overtraining Even at low intensity. That's if you do high intensity, you haven't active activity and how your body makes muscles. And muscles help make energy. And your pain response goes down. And you make more responses to your brain and oxygen content. And you can reduce your insulin sensitivity and increase your immune system. There is a greater risk for oxidative stress and a risk for overtraining, but I'm not asking you to do that. I'm not asking you to exercise 4 hours at a time. This is just going to be a few minutes. Now for those of you who have an autoimmune condition or have ever heard the term leaky gut, there is a role that exercise plays and the guts mitochondria during exercise. For instance, there's a cross talk between the gut mitochondrial and the gut microbiota meaning the bacteria in there during exercise. In essence the bacteria talk about making energy.

When you start to exercise it starts to communicate and has some other signals that increase oxygen uptake. This means they are going to have greater energy availability and better performance. Which means you're going to be resistant to being tired and you're going to increase your endurance. So, when you have a leaky gut and you choose not to exercise it's not good. How many doctors have told you
that?

HIT vs easy to moderate workout

Goal of workout	HIT	Jogging/biking/intervals
Increase immunity	Best	Little
Reduce nociception	Best	None
Muscle growth	Best	None
eNOS	Best	Best
Blood sugar/insulin	Best	Little
Brain protection	Best	Best
ROS and excessive training risk	Best	Little

Here's the other thing when you're trying to help your microbiome diversity; it's more than just a probiotic. It's moving into intense activity because the bacteria in your microbiome start to diversify, and they start to create short chain fatty acids that help protect your microbiome and your gut barriers. So yeah, if you have an autoimmune or inflammatory condition and you're not exercising can you see how you might be kicking your own butt?

There are nutrients that help support that mitochondria by energetics. They're minerals, vitamins, co-factors, fats, transporters, antioxidants, and herbals. Some examples are magnesium calcium phosphate for minerals. Vitamins C, D, A and B vitamins. Cofactors such as alpha lipoic acid and co Q 10. Transporters such as L-Carnitine. Good fats are known as essential fatty acids. Herbs such as Sukumaran and antioxidants such as glutathione and Resveratrol. For the most part we carry many of these in our online store because the literature and research has shown these to be effective for these conditions. There are many places to get them, and you have a source that you trust that's reliable and has research backward or sourcing you have a chance.

CHAPTER 6

Things That Absolutely Steal Energy and Recovery

Pain syndromes are very relevant to energy and there is an overlap between being tired or fatigued and pain syndromes and depression. When you have more fatigue, you have more pain. When you have more pain, you have more fatigue and often depression. We have more depression; you have more pain and fatigue. So, when you don't have any of these it's hard from the outside looking in because your only indicator is how you feel. Because you don't live in their shoes, you can't know how they feel. When they have these conditions going on, what stubbing your toe might feel to you might be amplified by two to 20 times more for people that have these conditions. The big picture here is mitochondrial dysfunction or reduced bioenergetics is more than just fatigue, pain, and depression right. It gets even more serious because when you have pain syndromes and fatigue syndromes and depressive and or mood disorders it then over acts and overlaps with neurodegeneration. If you're not familiar with the term neurodegeneration is something we all have. It starts about the age of 14 months and progresses through life. Now, when you're born you have the most amount of brain cells that you'll ever have. And while you have more brain cells than I do as an infant I can catch a ball and you can't as an infant. And the difference is I, and a developed adult, have neural connectivity and neural plasticity that allows me to function with complexity. And you just have a neuron that is just sitting there right. It is life that allows us to have the interconnectivity that allows us to do movement and complex solutions such as catching a ball. Now when you have these other syndromes it adds to your inability to have the interconnectivity and neuroplasticity which then exposes your weakness. For instance, if you were not good at math in grade school or high school but you've been able to manage as an adult when you start to degenerate it will show up in your math skills first. In order to make this better we have to do the things that you're not very good at because that's the part of the brain that needs to be worked out. In neurology to create neuroplasticity as we have to work on the things that we suck at.

Memory and Brain Function

Now that I talked about brain connectivity, I must talk about pain patterns. I want you to understand the perception that pain is brain-based, and that pain does not equal the amount of damage that you have. There are many physiological and psychological variables that can impact pain. Chronic pain is due to an activation of pain Pathways or a lack of activation or pain inhibition pathways or both. In our office we specifically focus on nociceptive fibers which are the pain pathway reduction. There are a multitude of ways to do this: putting your hands on somebody, using a 3B laser, microcurrent are other mechanical stimuli that are very good at helping the inhibitory pathways. Now, if you don't have enough energy and your cycle this

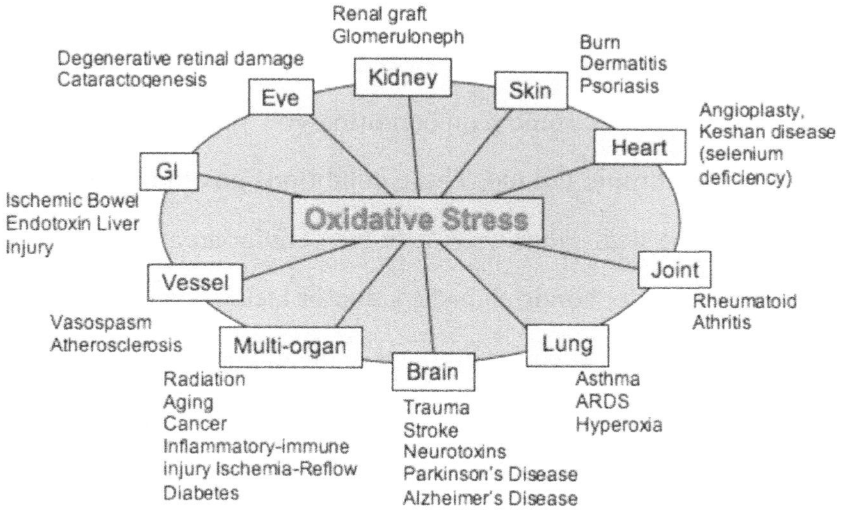

is how you can go to an acupuncturist, massage therapist or chiropractor and your pain is better for five minutes, an hour or a day and it comes back. We're somebody else might have the same treatment and they are great for a month. The more that we can stimulate these pathways the more efficient they can become. In our office our focus is diet, lifestyle, and nutrition. I expect my patients to exercise. I expect my patients to eat things that don't steal energy. So that when I do get to see them, and I do get to put my hands on them and use the tools that I have in my office, it's more efficient in creating neuroplasticity in their pain reduction centers and they get a better response.

Left the difference between when you feel pain when you don't or how you feel it versus somebody else. I want you to keep this in mind; if you have a chronic pain condition you also know you're going to have chronic fatigue. These conditions go alongside each other because it steals your energy. In the cellular area of your joints, you get mitochondrial dysfunction or lack of bioenergetics. This means the joint doesn't eliminate or clean up or repair. Which means it degenerates. Now you have inflammatory responses because it's breaking down and then that leads to pain. Here's where it gets more disturbing. When you have chronic pain, it can stimulate more chronic pain even though you're not doing anything specific to make it worse.

These conditions are called central sensitization and have been assessed and validated in the literature for pain/brain syndromes.

Here are a few conditions:
Fibromyalgia

Endometriosis

Vulvodynia

Chronic pancreatitis

Irritable bowel

Migraines

TMJ pain

Myofascial pain syndrome

Shoulder impingement syndrome

Multiple chemical sensitivities

Arthritis

Rheumatoid arthritis

Neurogenic pain

Whiplash

Post-traumatic stress disorder

Restless leg syndrome

Chronic fatigue syndrome

Interstitial cystitis

pelvic pain

dysmenorrhea otherwise known as menstrual cramps in pain.

In our office we have a sheet called RMP that stands for resting membrane potential. It explains how the nervous system works. The nervous system by itself is an all-or-none potentiation. Meaning it either fires or it doesn't, there is no in between. So, with the proper amount of oxygen, glucose, stimulation and lack of oxidative stress you get healthy nerve mitochondria. And when you get that you get the proper amounts of energy known as ATP end good nerve endurance. On the left side of the sheet things are good and you make great energy. On the right side, things are impaired and you make less energy, don't recover and are sick and tired and often full of inflammation and pain.

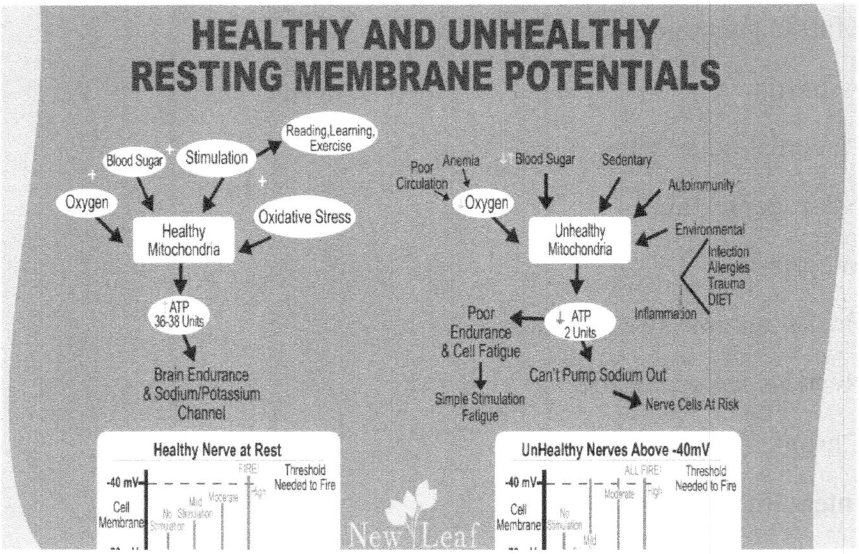

Blood Supply

When you have poor circulation or anemia it leads to a decrease in oxygen. When your blood sugar goes up and down it changes your glucose content which means no fuel to the cell. Or you have a lack of activity meaning in activity you have decreased stimulation. When you have inflammation whether that's food allergies or autoimmunity or environmental toxins or other disease States you get oxidative stress my way free radicals. All of these make unhealthy nerve mitochondria. All of these decrease in ATP production. And all of these create poor brain endurance. This is how you get fatigued and you're tired from stimulation or anything else you do. If you have pain and your inhibition pathways don't have the energy to dampen it because you have poor nerve endurance. Think of if you've ever stubbed your toe. When you first do it it's dadgum painful. And hopefully it doesn't break it but in the next 3 to 5 minutes that pain dramatically goes down. And hopefully it goes away by the end of the evening and by morning you don't remember anything happened. That's literally most of our injuries that we have. Now the bigger injuries when you really torn something or broke it they will continue to have pain. Let's say you don't have the ability for the nerve to dampen your pain pattern. Now all night or all day you have the same pain that you had when you stub your toe. So, it appears and it feels like you have the symptoms of breaking it. But now you go to the doctor and whether it's an X-Ray or an MRI or CT or an examination or however they do whatever they're testing or treating for whatever pain you have oh, nothing is found. Well, isn't that frustrating?

That is the gist of my practice. People who are frustrated, missing answers, and are sick and tired of being given the runaround. Most of my patients have been to multiple doctors with multiple chronic conditions whether it's pain or fatigue and they're told there's nothing they can do for him or her or it's just part of aging. I checked for this condition called “malingering" which is also a fancy term for faking it. And in the two decades of helping people solve pain and fatigue and over a hundred thousand visits, I've only seen one person where they were malingering. The reality is the practitioners that they saw before weren't checking for how the body makes energy. Then I have to ask, does the person in front of me make energy adequately or efficiently? Can it affect them differently than how we're taught in school? And the answer is yes. Once I had to send a patient to the emergency room because they had severe abdominal pain and symptoms were consistent with a burst appendix. They were, at first, untreated because they were treated only in pain. Upon questions they asked this patient on a scale one to 10 with 10 being the worst pain this said my pain is about a six or seven. The emergency room didn't think that was high enough to have a burst appendix and didn't do any further testing or vitals. What they failed to understand is that this individual was an Olympic athlete in track and field. This person turned themselves inside out in pain most days to get to the Olympics. Pain was a way of life. We sent them back with requirements for blood tests and thank goodness they did. It had burst and he was very ill, and he was able to get the treatment he needed to save his life.

Fatigue syndromes

The main group of fatigue syndromes are classified as chronic fatigue syndromes and fibromyalgia. Now in chronic fatigue syndrome you can't have a condition that could make you tired. But you have fatigue, sleep problems and brain fog or cognitive difficulties. And fibromyalgia you can have a condition and there are no medical exclusions for making the diagnosis of fibromyalgia. When you have no other cause for body white pain fibromyalgia is given to you as your diagnosis. But you can have a condition and you get diagnosed with secondary fibromyalgia due to that pain. The diagnosis for fibromyalgia requires at least three months of pain and tenderness in at least 11 of 18 areas. Never mind she also has fatigue, sleep problems, and cognitive difficulties. I bring this up because pain cannot be objectively evaluated. If you ever went to a hospital, one metric to decide what their efforts for you in managed skilled nursing is going to be is pain. However, your description of pain involves a cognitive function. So, if you have a lack of cognitive function from a chronic fatigue or inflammatory condition, then it becomes very difficult for you to explain how you feel. And someone who doesn't have this can only understand what you say. In other words, you may articulate to your nurse or pain doctor that your pain is a 3/10 but it really is a 9/10.

When it becomes fibromyalgia or a complex regional pain syndrome, there becomes a point when there is nothing from a stimulatory standpoint that's your trigger. In fact, you become the trigger. With pain sensitization syndromes you can have these conditions called phantom pain, or post-surgical pain, or from a traumatic brain injury. Those also self-replicate their own pain patterns without you doing anything. Again, you are the trigger because you don't have enough energy to slow it down. It takes energy along the PAG pathway to stop the sensation of pain.

And then there are active pain receptor stimulations. What this means is there's inflammation, degenerative joints, autoimmune disease, musculoskeletal conditions, blood supply and metabolic issues. These also include inflammatory bowel, endometriosis or congestive obstructive pulmonary disease, COPD. It can be osteoarthritis, degenerative disc disease, a rotator cut or torn meniscus or muscle. This could be migraines or peripheral artery disease. You can have diabetes or kidney disease. Or you can have autoimmune neuropathy, Crohn's, Celiac, rheumatoid, or other autoimmune conditions.

If you're sick and tired of being sick and tired and you have a pain condition it amplifies how you feel and, if you haven't seen anybody to look at those patterns or perhaps the metabolic origin of your energy your outcome is generally not very good. You're probably going from doctor to doctor, and you might hear something that makes sense but the application of it doesn't seem to help. For example, if you have pain and you have a sedentary lifestyle and you're obese here's what happens.

You get mitochondrial impairment meaning decreased energy. And then a decrease in oxygen-producing capacity and new neurons that start to fire more. This leads to increased information and increased insulin and increased fat cell size. Resulting in increased body weight. And when you decide that you're going to move, if any you get muscle recruitment because there's extra weight and then you get muscle pain. This movement then strains the ligaments, and you get pain. The new pain is called sclerotogenous pain. And now you have more inflammation within the ligament. On the anatomy cellular level, ligaments are surrounded by nerves which then fire into the pain pathways. That stinks. This is pain plus more pain.

And, if you're not sleeping, have blood sugar issues, anxiety, depression and stress, the pain feels more intense.

Let's assume person who is obese tries to exercise like you have been told for 30 minutes a day, five days a week. And they have a traumatic event such as falling or hitting their head. How are they going to recover? Are they going to get better or worse? These are what is called vicious cycles. They amplify each other and make things worse. For those in this vicious cycle you can't supplement or medicate your way out of this. You have to fix it on the cellular level.

Nutritionally there are good fats and bad fats and they come into a name called prostaglandins. If you ever heard of essential fatty acids that you can get from fish oil or healthy fats such as olive oil or avocado oil or flaxseed organic Meats they are anti-inflammatory. Better for eating those pesky grains with gluten an abundance of dairy or Halle amount of high fructose corn syrup again you can't supplement your way out of this. To make matters worse, many of the NSAIDs for pain have a rebound effect that amplifies by way of prostaglandins more pain because they decrease the good fats. Therefore, you get more information and more sensitization to pain. Please don't think that I'm trying to get you away from any of your over-the-counter medications. However, they have a limitation, and they are to be used short term when you need them. None of them are manufactured or designed or researched to be used every day for the rest of your life.

Some people have other triggers that lead to these pain patterns within the nervous system. And some people have histamine response and histamine comes from a cell in your body called a mass out. Now the idea of the mast cell is to reduce histamine to stop an invader from getting inside of the body or causing more issues. For acute issues, whether it's from histamine or other things, some of the best nutraceutical or supplemental products have to do with essential fatty acids because they reduce the inflammatory prostaglandins. Turmeric and Resveratrol decrease the expression of a molecule known as NF KB which creates inflammation at the cellular level. If it's a histamine response or another inflammatory molecule known as leukotrienes quercetin is one of the best nutraceuticals out there. But clinically what we see is the dose has to be between one and three grams, and most products have between 100 and 300 mg meaning you have to take 10 to 30 of those pills a day to get your dose. Also, there are things that reduce free radical activities such as NAC, glutathione and other antioxidants. When you have acute information if you can reduce your sugar and fast-food fats, shellfish, red meat, and processed foods while you're increasing your water, fatty fish such as salmon or tuna or mackerel and increase high antioxidant foods that are above ground vegetables it can help reduce your pain. Also, you can take cold showers or ice plunges or something called cryotherapy. These help increase endorphins which make you feel better and reduce pain, they reduce inflammation, they increase growth factors and the circulation through the end of the elite nitric oxide and here's the big and it increases your antioxidant enzymes. So just by using cold therapy you're going to help the mitochondria be more efficient.

As a reminder anything that increases inflammation will create more pain. And the longer that they're there the longer the pain pattern stays there, or your body can just be accustomed to it and self-perpetuate the pain itself. There could be an infection, or like a dietary protein such as gluten or corn or dairy, or autoimmunity or other forms of inflammation. This is how somebody can go gluten free whether they have a lab test that proves it or not and one person will feel much better, and the other person will feel nothing at all. If you went gluten-free for four weeks and you feel better this is great. If you went gluten-free and worse, we have a test that shows you are gluten sensitive and you feel nothing, or worse, we have a much bigger problem. It's not that these diets or dietary restrictions don't work, the problem is that practitioners or the lay individual do not understand what this means diagnostically. You are gluten free, and you have a gluten sensitivity and you felt nothing, or you felt worse. You potentially have a self-perpetuating pain pattern, or a self-perpetuating inflammation. Again, you're not going to supplement your way healthy or medicate your way to being healthy when this is occurring. You need a lot more.

One of the things you can do at home to increase your ability to slow down paint patterns is to increase your opioids naturally. There is an opiate crisis in the world, and I'm not suggesting taking any of those. I am suggesting creating your own opioids from creating endorphins and opioids directly. This activity can be physical exercise as we discussed before, laughter, healthy relationships, being touched, sex, volunteering or being in a pursuit of the passion that drives you, like yoga and meditation. When you increase your opioids, you activate the inhibition of pain.

Histamines

For those who have mast cell release of hormones your pain pattern is very difficult to reduce. Things that can trigger your histamine response are seasonal allergies, food allergies, environmental allergies, or infections. It could be the mast cells are like pain patterns and self-perpetuation and activating themselves. These individuals absolutely have to load up on antioxidants and we do our best to quench these free radicals. They reduced the histamine with something called quercetin. Again, at the fundamental cellular level we have to get these individuals to make more energy through ketosis and then to help the mitochondria by way of intermittent fasting.

Impaired gut function

I have to say something about a condition called a leaky gut, a condition I have mentioned a few times already and is vital in proper management to make energy. Clinically it's called a permeable membrane. And before you get started or think this is just a movement, there are 300 published research papers on permeable membranes. It's a real condition and that leads to a myriad of diseases including autoimmune disease, cancer, heart disease and neurological conditions such as Parkinson's, Alzheimer's, and dementia. Why do we pay attention to the potential of leaky gut at least in our office? Now you might not be in our office so what can you do to prevent a leaky gut, and what are the causes of leaky gut.? There are several different pathways I will go through. If you recall you can have metabolic issues such as diabetic or an autoimmune condition. You have inflammation inside of your intestines and that could be from food, or you can have infections of any type that can break down the barriers. You could be going through a stress response whether it's work or relationships or you're not sleeping which means it increases the stress response. You can have hormonal imbalances whether that's from your thyroid because it regulates your digestion, or you're going through puberty or menopause or cyclic dysfunctions. You can have brain issues such as degeneration that then makes your leaky gut worse and vice versa. You can have a stroke or a traumatic brain event or a concussion that can create a leaky gut. And of course, there's dietary concerns whether it's gluten or dairy, processed foods, excess sugar, fast food, high fructose corn syrup and even alcohol can be a cause of leaky gut. And, antibiotics and antacids steroids and other medications, although given the help you, can also trigger and perpetuate a leaky gut.

To help manage a leaky gut nutritionally you really need to have vitamin D or bone broth real glutamine and this thing called short chain fatty acids to help it heal. Often doctors are quick to reach for probiotics and digestive enzymes which can be helpful but that's not the first line of defense. Typically, they come later but if that's working for you right now that when you take down you feel a whole lot better by change?

CHAPTER 7

Food Sensitivities (Allergies) and What To Eat For Your Health

For the most part, we have a Standard American Diet, SAD for short. We are 37th in world health, yet for decades, we have been first in spending. That's not the number one we should be focused on. Spending is not prevention. I've decided to write down a few highlights to get everyone started. Because we live in a toxic world, 350 different pesticides, 75,000 chemicals utilized in industry, new carpets, xenohormones and on and on, it's obvious we can't avoid being exposed. I like my carpet where I watch football games. However, we can take efforts by being responsible for our environments and what goes into and on our bodies. I will go into everyday toxins in the next chapter because it does affect digestion. DIE-T is a four-letter word that doesn't work. Lifestyle modification works. I promise you that you can do anything if you're big enough. You may have to modify, change, adapt and integrate any plan. You can do it.

Food allergies and sensitivities have become much more common in today's world. They are very complex and vastly overlooked when a health problem arises. The most common way to test for food allergies is with a blood test followed by an elimination diet. Certain questionable allergic foods are eliminated from the diet for 3-9 months. Then they are reintroduced. This is easier said than done.

Many people are allergic to their favorite foods, and they crave them because that cannot absorb them. Often allergies are created by structural disturbances and become neurological. Some food allergies are due to food being introduced in infancy instead of one at a time. As an infant, or if you are introducing food to a small child coming off breastfeeding or formula, one should introduce one to three foods a week separate from other foods.

The microbiome of your child and of you consists of all the cells that are not human in nature. This means in the womb and during natural birth. A child's microbiome is inhibited by cesarean. The microbiome is 50-55% of your body weight from cells that are symbiotic with us. It is made of bacteria, yeast, fungi, and parasites. Gains in strength and tolerance are noted in infants when given foods one at a time to identify, metabolize, and dampen inflammation. When we have patients with a lot of microbiome issues, we start them with a very diverse shake to help decrease the inflammation. The plan is so they can absorb the food and nutrition that we are suggesting for their case. It's also for diversity which will help their microbiome.

Food allergy testing can come is several forms, skin scratch or blood for IgE. These are very serious allergies. Blood for IgA IgG for systemic inflammation. Once you have a positive finding avoidance of these inflammatory foods allows the body to clear the immune properties. When you begin to reintroduce a suspected food or foods, do it one food at a time. Eat that food throughout the day along with non-allergic foods, paying attention to any reactions. This could be headaches, dizziness, nausea, or other symptoms you've had that have gone away or decreased during the three-week period. If you have a reaction, you need to avoid it for 60-90 days. Wait three days before reintroducing the next food if you have more than one sensitivity or questioned food.

The most common are wheat, gluten, milk, peanuts, corn, and milk. Soy, sugar, citrus, shellfish, coffee, and food additives are second. Look for replacement foods before beginning or you may feel starved, even though you're not. Replacements can help you cope with the loss of something once savored.

If you're not doing it already, limiting gluten, dairy, sugar, soy, eggs, corn, nightshade vegetables, alcohol, and refined foods, this is when you should start. I know it sounds like a lot. You can eat as many dark, leafy green vegetables other vegetables and fruit in the beginning. You can have nuts and seeds and healthy sources of animal protein, which means they are free range grass and grass-fed, antibiotic free. If this doesn't work, we start taking some of these away when you are autoimmune or have severe digestive issues.

If you are a sugar addict, sticking this out for three weeks will help with your sugar cravings. You might also find that symptoms such as bloating, sinus congestion, and excess gas also reduce or go away. Introduce one of these inflammatory foods once you've passed your two or three weeks or none. Observe symptoms, and in three days, introduce another. Let's say you ate corn on the 15th day and all of a sudden you bloat or swell, and you haven't been doing that for a little bit. That's a great indicator that corn is a problem. You'll need to wait three days or more until the swelling is gone before you introduce another food.

Charting your foods is recommended so you know what is good and what isn't, and when you can eat them again, if ever. If a method has failed, a food allergy sensitivity test can be ordered through your doctor or on our website. However, this is buyer beware regarding food sensitivity testing. Most food sensitivity companies test only raw food. My question is, how many people eat raw beef, chicken, or fish? Not many. There's a difference between how your body reacts or can react to raw or cooked food. If and when you are tested, you need both of these scenarios/options tested.

Furthermore, one may not be allergic in food allergy testing to wheat, tomato, or cheese individually. But combine them together in something called a pizza, it changes the expression of the proteins. And yes, one can indeed be allergic to pizza and not the individual ingredients. Frustrating, yes. But a doctor who understands this can help you get there. The gold standard for any food allergy testing is Elisa blood serum testing. Skin scratch and cytotoxic testing are hard to reproduce, and in my humble opinion, a waste of good money. You can have an Elisa blood spot test which is better for children.

SPECIFIC Food programs, some you know, some may be new to you to decrease inflammation and make energy:

Let's start with physiology. There is a difference between hunger and cravings. If you have cravings, you are missing nutrients. It's hard to not turn your brain off when you are "joshing" for something. Do your best to get it out of your head. Eat something healthy, asking yourself if you need to eat it or do you want to eat it. Honesty goes a long way here as we can justify just about anything in our own minds.

FODMAPS

For those who have bad intestinal issues or struggle with irritable or inflammatory bowels, you have an autoimmune bowel. You may consider the FODMAPs for a few months. The term 'FODMAP' is an acronym for Fermentable Oligo-, Di-, Mono-saccharides, and Polyols. In short, they are fermentable carbohydrates found in many of the foods we often eat. If you are on FODMAPs or considering it, you should consider working with a professional. Also get tested for permeable membrane and bowel infections so you know how to help yourself.

Autoimmune Paleo

Many of our autoimmune patients are on the autoimmune Paleo diet (AIP) which looks exactly like elimination without nuts. The goal is to restore your microbiome. You do this by eating as much variety as possible. If your symptoms are going from an 8/10 and to a 5/10 or lower, you know that this is working. But you also know this doesn't happen overnight.[3] [4] [5]

Ketogenic

The ketogenic diet has been very successful with those who have brain disorders such as epilepsy or seizures. There's a brain modulator when you're in ketogenesis called beta hydroxybutyrate made in the liver. While in keto, this helps create energy in the brain. It doesn't matter how you get there, whether it's plant-based protein or animal-based protein. Beta hydroxybutyrate affects the immune system and helps support certain regions of the brain and genetic expression of BDNF. A proper ketogenic diet consists of a lot of fibers. Particularly green-leaf vegetables. The ones that have the highest amount of fiber are best. Swiss chard, bok choy, spinach, kale, and arugula with some sort of protein combined with it. In a perfect keto program, you cycle through four weeks with measurements. You still need blood tests before and after with measurements of urine every three to seven days to make sure that you are in ketosis. For menopausal women who started gaining weight with menopause, or women gaining fat on the back of their arms which is very hard loose, the term for this is estrogen dominance. Keto is beneficial to reducing estrogen dominance and it supports female physiology.

The caveats and warnings come before going on a keto diet. You need to make sure that you can digest food first, especially fat, because that's been your macronutrient. So, if you don't digest fat or have a gallbladder issue, your stool floats. If you have low vitamin D or low cholesterol, the ketogenic diet isn't going to work very well for you. And if you still want to, you need to fix fat digestion first.

Intermittent Fasting

You can start intermittent fasting (IF) or consider it as an option as long as you are not hypoglycemic. With intermittent fasting you need to be fasting for at least 12 hours minimum with 16 hours being the max. With intermittent fasting, as the day goes on, you have a better response to insulin. But those who are hypoglycemic doing intermittent fasting now get worse because you're enhancing your hypoglycemia. You must unwind hypoglycemia first. Once you manage your blood sugar and urinalysis proves you are in ketosis, now you can expect the benefits of IF. You want to skip your breakfast because it gives you more time to fast, but every other day. The result is your blood sugar spikes being stabilized because insulin responds faster and is more efficient. What intermittent fasting (IF) does is it kicks in your metabolism. Sometimes IF gets combined with low calorie intermittent fasting. That's all relative to where you are in your health. If you have any factors that would make this not safe, don't do IF. In our office we use several programs to get someone into ketogenesis. Either rapidly by a fast mimicking one-week program Prolon program, or a long-term anti-inflammatory ketogenic program. Sometimes you need one or the other or both. I prefer using these before going full blown into ketogenic or IF. But IF can be combined with the anti-inflammatory ketogenic program to maximize your chemistry. [6] [7] [8]

CHAPTER 8

The Toxins Around Us and Between Your Ears

We are exposed to a lot of toxins in our world. Toxins come in many different forms. Most of us are aware of pesticides, heavy metals, herbicides, fungicides, and industrial cleaning solvents. But toxins also come in foods, especially GMO. They come in the air we breathe, the carpets, paint, and cleaners of our homes. Or, when we drive our cars, enter a building that is being refinished or an old building with asbestos. There are also plants that are toxic to touch and what we breathe from them. There are even toxic relationships and thoughts. The bottom line is, we need not live in fear of being exposed to toxins. You are exposed on a daily basis. You should try to limit exposure when you can so that your body can express its potential.

Not having a good removal system of waste can affect how you sleep or create insomnia every now and then. That the true measure of your fluctuation. If we cannot detoxify, we have imbalance in hormones, fatigue, and weight gain. Toxins also increase temperature. That said, the people who tend to have the most insomnia are often functional at a higher temperature. Your body drops its temperature at night when you sleep. This helps to release sleep-related neurotransmitters. These turn on sleep related hormones so that you can sleep. Should you have trouble sleeping, you might consider a cool pillow or cooling sheets, or even lower the temperature in your home before you go to bed. It's an easy start.

For toxic chemicals in the body, you get glutathione depletion. You can get a loss of self-tolerance. A reduction in T-regulatory cells. An increase in autoreactive cells and attack your own DNA. You can have chemicals bind to proteins and become chemical sensitivity. Toxins increase oxidative stress, and they decrease mitochondria function thus decreasing energy. They add to intestinal, lung, and brain permeability. In other words, leakiness. You need to do some necessary steps to reduce them. Using an infrared sauna with all three wavelengths is a great start because you can't avoid everything.[9] [10] [11]

There are a few definitions that will help you understand toxins better. Dose-dependent toxicity is what a lot of manufactures use to label a chemical or drug "safe." This means that for the dose or exposure recommended, there appear to be no adverse health risks. Dose-dependent, as in the case of caustic chemicals, are labeled harsh as one dose can cause inevitable harm without protection.

The other option that most people are unaware of is a build-up model of toxicity. Over time, a low dose quantity is considered very safe. After continued exposure, the chemical is increasing its concentration into your body. This can build up to a significant quantity that is now harmful to your body and is a toxic dose. This dose can lead to reactions and illness even though you were careful and using the "safe" recommended dose. These doses can affect hormone levels, thyroid function, and autoimmune conditions. [12] [13] [14] [15] [16]

There are a lot of tests for exposures, heavy metals, hair samples, urine, and blood. Keep in mind everyone has exposure to heavy metals. We live in an industrialized society. What does this mean? Nothing. Finding heavy metals in hair or other tissues is not something to be concerned with. However, if we find an antibody to the metal now, we have a serious finding. This is when detoxification and supporting your detoxifying organs, skin, kidneys and liver, are vitally important. A side note. IV Chelation, especially in those who are autoimmune is not warranted. In fact, it makes a lot of those patients worse.

CHAPTER 9

GOALS FOR MAKING ENERGY

Goals for energy production

A) *Goals of exercise for fatigue*

Be able to exercise and intensity to fully recover the same day to any workout

Exercising and intensity to avoid pain and fatigue

Increase over time but make sure you don't over train

When you have fatigue, it is imperative to make sure that you do not engage in exercises that overstimulate or fatigue out your nervous system and brain. These exercises include tracking balls, balance and coordination and thinking while moving. Do you remember back in the day I used to run and think about life and solve mathematical problems at the speed of which I was moving? This is not a great idea when you are low on energy because you're using energy to calculate those thoughts. And so, when you're trying to get movement it will bring up your energy and you don't get to exercise for the same duration and it takes longer for you to make more energy.

Before you exercise you should consider drinking some water or taking some glutathione because of the antioxidant production and also consider products that have vinpocetine and huperzine because they help increase blood supply. Now after workout you may want to consider a cold shower or cryotherapy.

B) *Goals for sleep*

I want you to sleep efficiently because it helps to remove unhealthy mitochondria. It helps to clean up any unhealthy cells by way of autophagy that happens during sleep. It helps increase melatonin and anabolic hormones that help your body repair. More importantly, it has a positive effect on mitochondrial biogenesis meaning more energy. This happens by way of a release of growth factors that supports plasticity and your brain and in your muscle memory. And it also helps convert what you are currently learning known as working memory to long-term memory. Make sure your blood sugar is correct, you are digesting, and you have added all the strategies from above for assuring quality sleep. You may need to consider stress reduction a few pages further as an actionable step to help you relax.

C) *Goals for pain management and inflammation*

I really want you to sleep in order to manage pain because there are a multitude of studies linking sleep and pain perception. The journal neuroscience in March of 2019 explains that the activity in the sensory part of your brain increases by 126% when you have a sleepless night. When we are talking about pain management strategies, sleep is extremely vital. Also, if you don't sleep you increase your sympathetic nervous system, so you are in fight or flight mode. This increases your NFKB (inflammation) signals which then creates more inflammation throughout your body. And chronic inflammation then leads to low energy production and depression. And then of course depression doesn't allow you to sleep well, and the cycle continues.

The key concepts are to find out where you are missing your ability to sleep. Are you able to fall asleep, stay asleep or both? And are you able to get up and feel recovered when you do sleep?

End for chronic pain, you have to manage your blood sugar meaning that you can't have foods that create insulin surges, and you can't be missing a lot of meals until you've gone into ketosis then intermittent fasting to where you have the proper amount of blood sugar, and you don't have hypoglycemia symptoms. If you are uncertain of the symptoms, review above the symptoms of hypoglycemia and hyperglycemia related to insulin resistance. You need to not combat those issues while you're trying to make energy. Pain dysfunction of blood sugar also sensitizes pain receptors in your body, and you get increased pain patterns. Thus, you need to stabilize blood sugar before you can focus on pain patterns.

The last thing is if you have a pain pattern that's affecting how your body makes energy and you are just trying to supplement or perhaps exercise or medicate your way out of it you are so very lost and I feel for you. In cases like this there is a neurological term that is called cortical mapping. In the world of cortical mapping, if I touch your toe it is going to go to a particular part of your brain that is related to your toe and your toe has more sensory than let's say the outside of your hip. There is a neurological map called the homunculus of man. This map shows that you have a lot of sensory receptors in your lips, fingers, eyes and feet. Compared to your knee or elbow, where you don't feel much sensation. When you have pain patterns you have to stimulate an offsetting pathway called the periaqueductal grey (PAG) this is what slows down your pain patterns. When you get hurt eventually that PAG turns on and slows down your intensity of pain. Hands down through thousands of research articles, the best way to help this increase the PAG is with cortical mapping. If you're lucky and you have a pain pattern cortical map that's related to your spine, chiropractic works. If you have muscular cortical mapping perhaps massage and physical therapy will work. If you have neurological circuits perhaps electrical stimulation, be a microcurrent or TENS unit and acupuncture works. Or you may have a combination of all of these and you need to use them all or combine them all in a session. There are doctors who do this of which I am one and I am not alone. This is how some people feel relief with one doctor and therapy while others do not.

D) *Goals for food management for sick and tired of being sick and tired.*

There are seemingly an unlimited number of diets that you can adhere to when it comes to managing your food intake. Here is where we typically start in our office. First off, if somebody can go an Autoimmune Paleo (AIP) which is a little bit more comprehensive than just a Paleo or caveman diet, then they are taking a lot of the inflammatory based foods out of the system. Whole 30 does a good job, SHAPE Reclaimed does as well. All of these are great starts. What you're going to do is eliminate unhealthy fats such as fried foods or foods that have partially hydrogenated fats. When you are eating protein, you want to limit your red meats, and when you do eat them make sure that they are grass-fed and grass-finished. Also be very careful with shellfish when you have inflammation because all forms increase the inflammatory processes in the body. There are also foods that age you and they're called advanced glycolytic end products (AGE's.) Let me explain. When your food is caramelized or broiled or grilled or fried, it creates he's in products and they create inflammation. And there's foods that have just too much sugar such as white rice, breads, pasta, candy, and other desserts that create an insulin surge. These insulin surges lead to inflammation and preventing and reducing these are really good starting points that slow your inflammation down through diet.

We want healthy fats such as avocados or fatty fish or olive oil. And that we want foods that are high in plant fiber that grow above ground. The good fats will help decrease inflammation in the microbiome by supporting high fiber foods. These help decrease inflammation because they support your good bacteria and they are anti-inflammatory. And then there are superfoods such as berries, green tea, garlic, pomegranate, and acai berries that are high in antioxidants, which decrease inflammation.

The most common reactions that create histamine responses in the body are dairy, gluten, eggs, soy, and corn. Also, leftovers create more histamines so if you know that you have IgE based reactivity then you will have pain fibers that are also triggered when your mast cells create histamine. Again, these are really good areas to start.

If you haven't heard, I really do want people to go on a ketogenic diet for making energy, being able to sleep, being able to regulate blood sugar handling, and being able to reduce pain. But that’s not for everyone and you certainly know when it isn’t by how you feel. What a ketogenic diet does is reduce insulin response. This creates less response from your fat cells, less activated pain receptors, and less AGE’s that age you. You can increase ketones that will dampen microglia neuroinflammation. This means your brain is not inflamed. And that you get more energy for your mitochondria by way of ATP. That when you have more ATP, then you do not have your nerve cells ready to fire all the time and you're more efficient in your thought process and you can slow down anxiety and depression through this process alone. There may be other factors, but this is a fantastic physiological and fundamental start. Increased ketones mean increased brain function and reduced pain. You also have a decreased fight or flight response and more energy for you to use for other functions.

E) *Goals for a permeable membrane and immune system*

Gut support is also important because that is where the immune system assimilates. If you have a condition known as a leaky gut or have been told you have one, there is a lot more information than having a leaky gut. There are four different types of leaky gut and multiple factors that cause a leaky gut. It's best to know what type you have. In medical terms, a leaky gut is known as a permeable membrane. You need to understand the factors that trigger your leaky gut so that you can do your best to mitigate them. Those triggers and factors will be presented in another section, but it is warranted to work with somebody who understands these processes and can evaluate and test your triggers. You want a product that connects the membrane fibers and tightens up the gap junctions. These are the proteins that are broken down or lost, causing a permeable membrane. Your supplement needs research and publications regarding gap junctions. Otherwise, it's a high probability it has no value.

If you have a leaky gut, autoimmune disease begins in the gut with the breakdown between your small intestine and large intestine. This is where the majority of your immune system lies. If you do have a leaky gut or permeable membrane, it is important to understand what type. Are you paracellular or transcellular, do you or do you not have endotoxemia? Have you had traumatic brain injuries, or concussions? All these add to or take away from your outcome, and recovery.

There are multiple ways to address autoimmune conditions. Conventional treatment consists of medication and that's about it. It's about management and there is no true long-term solution other than increasing your dosage. For most people with autoimmune conditions, or who think they may have one, this is not your only choice. Those whose tissues are destroyed by their autoimmune disease may never recover. But they can have a better outlook on life by making some important changes.

You may have a leaking gut, otherwise known in medical literature as intestinal permeability. Over 200 clinical papers have been published on this topic on PubMed. A leaky gut can be a hindrance to recovery or a precursor to autoimmunity. That means it is a real thing, not something made up by alternative doctors. Our office uses laboratory testing to show how long your treatment plan should be and how to address this inflammatory condition. It is also to confirm an autoimmune condition and/or leaky gut and what stage you are in.

Below are just a few of the ways you can create intestinal permeability, but these are not the only options. There are more. Emotional (EMO) stress, genetically modified foods (GMO,) traumatic brain injuries (TBI,) high blood sugar (AGE's,), autoimmunity (AI,) neurodegeneration (ND,) and gastrointestinal (GI) inflammation. And how many ways are there to acquire these? The ways appear to be infinite.

In the case of gluten sensitivity, a main mechanism is the DPP IV enzyme. This enzyme is involved in the degradation of proteins such as gliadin and casein, a product of milk. DPP IV helps support the immune system. Also, you need brush border enzymes such as amylase, invertase, and lactase. These are your main gut enzymes. They support the digestion of carbohydrates, proteins, and fat without causing irritation and digestion of the intestinal walls. Without the brush border, you have intestinal permeability, or more well-known on the internet as leaky gut. Finally, key antioxidants have been shown to help intestinal response. These are glutathione and fat-soluble vitamins A and D.

When we have intestinal permeability, as well as several other conditions, we can lose our tolerance to many things. Often, out of the blue, we become intolerant to smells, metals on our body, shampoo, lotions, detergents, and foods. This leads to skin outbreaks and lots and lots of frustration. This is how someone can do a heavy metal hair analysis and it comes back positive. After treatment, those folks may not be better, or worse, they are indeed worse. Many doctors tell them they must get worse to get better, or they're having a healing crisis. I'm alternative minded, but this is absolute nonsense. Or there's something no one understands, and this is a new reaction. It could be a physiological change that was not investigated or understood.

Gut permeability has other causes than eating too much sugar or having a love-hate relationship with another person. It could be from infections or other sources of stress, such as brain trauma and even body trauma, or chronic sources of inflammation such as poor sleep and poor sleep habits. Or, malnutrition can lead to a loss of the immune system, which in itself has a limited tolerance for chemicals that we are exposed to on a daily basis.

Once we have lost our tolerance, we are either depleted in our glutathione, or we can no longer recycle it, leading to an increase or build-up of toxicity. The cellular function of our immune cells, especially those known as T-cells, are no longer regulated. Without this, we get inflammatory hormones running rampant. This is the beginning of food and chemical sensitivities. If you have deteriorated to this point, you are far down the road. It's not an overnight fix unless it recently occurred. Sometimes this happens with exposure to high doses of mercury or radiation. If the damage isn't too great, you may recover quickly. But that is rare. Once your T-cells are unregulated, the symptoms come. These are fatigue, persistent pain from inflammation, tissue breakdown and degeneration. You may even have a long or chronic illness, and even, dare I say it, autoimmunity. This condition is also known as toxic induced loss of tolerance or (TILT.)

F) *Quality-Stress control*

The fifth factor outside of supplements to boost your immune system is to live and enjoy life. When you're having fun through laughter and entertainment, you gain an opioid response. Opioids activate your T cells which improve immunity. This can happen through telling jokes, reading jokes, watching a funny movie, or catching up with a friend who makes you laugh.

Also, crying does have an opioid response so if you have an emotional event that you've been harboring are having trouble working through, let it go. If you can't do it on your own, please get some professional help and talk it out.

Talking about an issue brings out an opioid response which has a trigger to your brain to enhance your immune system. Holding issues in triggers a stress molecule and can slow down your immune system.

Emotional Freedom Technique

You can find many a funny chuckle online with our interactive world, but not always is there stress reduction or emotional health online. In the meantime, I am going to provide you a link for a tapping technique called emotional freedom technique (EFT). This is not a replacement for proper care or counseling, but we know you can't go to a counselor every day. You can do EFT every day. Link is (https://www.emofree.com/)

For those of you here sitting around watching TV and waiting for the world to implode, I highly suggested using EFT. Again, if you are constantly thinking about a situation that is causing anxiety, these will decrease your opioids. A loss of opioids is not what you're shooting for.

Limit yourself on what you watch. Personally, I spend no more than 10 minutes on the news per week. Most of my time is research looking at clinical trials and breakthroughs so that I can bring that information to my clients. That makes me happy. The rest of my time, I am looking for a solution we can do to prevent and decrease the negative effects of this pandemic and so much more.

Lastly, the right breathing stimulates your calming side of your nervous system and helps to improve your mental state. The bigger picture is that it will help boost and support the immune system. In general, there are four steps to help with proper breathing, so let's get started. Take a deep breath through your nose, expanding your belly and not your chest or shoulders. Breathe out slower than you typically do but not to the point of exertion, and exhaustion or discomfort. First, take a deep breath through your nose with your lips closed. Try to breathe in as deep as you can. Then, put your tongue at the top of your mouth and purse your lips together and breath out. When you do this for 5-10 breaths and with deep breaths, you can stimulate the lower parts of your lungs that have nerve endings to the parasympathetic nervous system and help relax you. Do this each day or more if you have many stressful events or live in a stressful environment.

Allergies/Histamine support

Histamine support may be required if you have systemic allergies, as allergies are inflammatory and breakdown barriers. Histamine causes in the intestines an influx of water, which leads to chronic diarrhea or loose stools. In the sinuses, you get drainage and post-nasal drip that can lead to a cough. While histamine is important, not managing it through diet and environmental awareness allows histamine to be inflammatory and bind up nutrients. These nutrients can be used in other parts of your immune system, like those that are necessary for fighting an infection. This is why I try to help as much as I can those who have allergies of any type. The bigger picture is they will have immune deficiencies elsewhere. Allergies in adults genetically makes sense. If you are over 50 years old and have a few genes that could be getting less efficient, and now you have allergies, that's normal.

Allergies in children are not normal. Their immune system is being challenged and that needs to be addressed. Sometimes addressing with medications is necessary. Understand this. If you suppress the immune system, histamine, and don't attempt to address the why, there will be more medications and more allergies as they grow. There is growing research that most medications for allergies decrease the blood-brain barrier and over time cause a leaky brain. By taking these at an early age, it increases brain breakdown, inflammation, and digestive disorders. Long term use of over-the-counter antihistamines is very scary. In fact, for every year you take them, the probability of dementia goes up.[17] [18] [19]

In recap, before you reach for a whole bunch of supplements, make sure that you have adequate sleep and you are hydrating enough. Enjoy what you can, work on yourself, move your body and exercise. And of course, you know this, do not binge watch TV, stay up too late, and eat whatever.

In closing, we talked about the methods of why you probably are sick and tired and why you are sick and tired. I have provided you some options to help unwind where you are and what condition you are in. Hopefully this at least gets you started to make energy. I have provided some dietary, exercise, and nutritional options on what to take. If you are still lost at what to do, I encourage you to take our survey to get a broad-spectrum view of where you should start. And lastly, I am going to give you a word of encouragement. You can do anything if you're big enough. If you haven't had success yet it is because the right pathway hasn't been open to you. For each of these ideas above, do it for four weeks and see what happens. Because what happens, good or bad, is diagnostic. Then your results can tell you what else you need to do. Or better yet, perhaps this is exactly what you need to do.

All my best ~ Dr. Trites

RESOURCES

To learn more about our office contact us at

office@choosenewleaf.com or call 469-491-7535

Office Funnel Link

Or go to www.choosenewleaf.com

Acupuncture resources including acugraph from Miridiatech

https://bit.ly/2EKFAsK

https://bit.ly/34ICD6Y

Emotional Freedom Technique

https://www.emofree.com/

Work out at home

https://8fit.com/dralantrites

Sauna

https://www.sunlighten.com

REFERENCES

[1] "Measuring Overall Health System Performance for 191 ...." https://www.who.int/healthinfo/paper30.pdf?ua=1. Accessed 8 Jul. 2020.

[2] "Major Depression and Activation of the ... - PubMed." https://pubmed.ncbi.nlm.nih.gov/10442165/. Accessed 8 Jul. 2020.

[3] "Paleolithic Diets as a Model for Prevention and Treatment of ...." https://pubmed.ncbi.nlm.nih.gov/22262579/. Accessed 10 Jul. 2020.

[4] "Paleolithic Diet - StatPearls - NCBI Bookshelf." https://www.ncbi.nlm.nih.gov/books/NBK482457/. Accessed 10 Jul. 2020.

[5] "Efficacy of the Autoimmune Protocol Diet for Inflammatory ...." 29 Aug. 2017, https://www.ncbi.nlm.nih.gov/pmc/articles/PMC5647120/. Accessed 10 Jul. 2020.

[6] "A Periodic Diet That Mimics Fasting Promotes Multi-System ...." 18 Jun. 2015, https://pubmed.ncbi.nlm.nih.gov/26094889/. Accessed 10 Jul. 2020.

[7] "Fasting-mimicking diet promotes Ngn3-driven β-cell ...." https://www.ncbi.nlm.nih.gov/pmc/articles/PMC5357144/. Accessed 10 Jul. 2020.

[8] "Effects of Intermittent Fasting on Health, Aging, and Disease ...." 26 Dec. 2019, https://www.nejm.org/doi/full/10.1056/nejmra1905136. Accessed 10 Jul. 2020.

Chapter 10

[9] "Repeated Sauna Therapy Attenuates Ventricular Remodeling ...." https://pubmed.ncbi.nlm.nih.gov/21622828/. Accessed 10 Jul. 2020.

[10] "Sauna Detoxification - Infrared Sauna Detox Benefits ...." 19 Nov. 2009, https://www.sunlighten.com/infrared-sauna-health-benefits/detoxification/. Accessed 10 Jul. 2020.

[11] "Clinical Effects of Regular Dry Sauna Bathing: A Systematic ...." 24 Apr. 2018, https://www.ncbi.nlm.nih.gov/pmc/articles/PMC5941775/. Accessed 10 Jul. 2020.

[12] "A study of thyroidal response to thyrotropin (TSH) in ...." https://www.ncbi.nlm.nih.gov/pubmed/2484874. Accessed 10 Jul. 2020.

[13] "[The influence of treatment with substitutive or suppressive ...." https://www.researchgate.net/publication/7259178_The_influence_of_treatment_with_substitutive_or_suppressive_doses_of_thyroxine_on_biochemical_bone_turnover_markers. Accessed 10 Jul. 2020.

[14] "Change of Bone Mineral Density and Biochemical Markers of ...." 31 Aug. 2015, https://www.ncbi.nlm.nih.gov/pmc/articles/PMC4572035/. Accessed 10 Jul. 2020.

[15] "Removal of Dental Amalgam Decreases anti-TPO and anti-Tg ...." https://pubmed.ncbi.nlm.nih.gov/16804512/. Accessed 10 Jul. 2020.

[16] "Phenotypic Dichotomy Following Developmental Exposure to ...." 25 May. 2009, https://pubmed.ncbi.nlm.nih.gov/19433254/. Accessed 10 Jul. 2020.

[17] "Can Allergy Medications Harm Your Brain? | Cognitive Vitality ...." 25 Apr. 2018, https://www.alzdiscovery.org/cognitive-vitality/blog/can-allergy-medications-harm-your-brain. Accessed 8 Jul. 2020.

[18] "Common anticholinergic drugs like Benadryl linked to ...." 25 Jun. 2019, https://www.health.harvard.edu/blog/common-anticholinergic-drugs-like-benadryl-linked-increased-dementia-risk-201501287667. Accessed 8 Jul. 2020.

[19] "Long-term Anti-inflammatory and Antihistamine Medication ...." https://www.ncbi.nlm.nih.gov/pmc/articles/PMC6436627/. Accessed 8 Jul. 2020.

Made in the USA
Monee, IL
04 November 2026

67381316R00066